ENGINEERING MATERIALS

Applied Research and Evaluation Methods

ENGINEERING MATERIALS
Applied Research and Evaluation Methods

Edited by
Ali Pourhashemi, PhD

Gennady E. Zaikov, DSc, and A. K. Haghi, PhD
Reviewers and Advisory Board Members

Apple Academic Press
TORONTO   NEW JERSEY

Apple Academic Press Inc. | Apple Academic Press Inc.
3333 Mistwell Crescent | 9 Spinnaker Way
Oakville, ON L6L 0A2 | Waretown, NJ 08758
Canada | USA

©2015 by Apple Academic Press, Inc.

First issued in paperback 2021

Exclusive worldwide distribution by CRC Press, a member of Taylor & Francis Group
No claim to original U.S. Government works

ISBN 13: 978-1-77463-359-5 (pbk)
ISBN 13: 978-1-77188-043-5 (hbk)

Library of Congress Control Number: 2014950947

Library and Archives Canada Cataloguing in Publication

Engineering materials : applied research and evaluation methods/edited by Ali Pourhashemi, PhD; Gennady E. Zaikov, DSc, and A. K. Haghi, PhD, Reviewers and Advisory Board Members.

(Advances in hospitality and tourism book series)
Includes bibliographical references and index.
ISBN 978-1-77188-043-5 (bound)
1. Materials--Research. 2. Materials--Evaluation--Methodology. I. Pourhashemi, Ali, editor

TA404.2.E55 2014   620.1'1   C2014-906439-X

Apple Academic Press also publishes its books in a variety of electronic formats. Some content that appears in print may not be available in electronic format. For information about Apple Academic Press products, visit our website at **www.appleacademicpress.com** and the CRC Press website at **www.crcpress.com**

ABOUT THE EDITOR

Ali Pourhashemi, PhD

Ali Pourhashemi, PhD, is currently a professor of chemical and biochemical engineering at Christian Brothers University (CBU) in Memphis, Tennessee. He was formerly the department chair at CBU and also taught at Howard University in Washington, DC. He taught various courses in chemical engineering, and his main area has been teaching the capstone process design as well as supervising industrial internship projects. He is a member of several professional organizations, including the *American Institute of Chemical Engineers*. He is on the international editorial review board of the *International Journal of Chemoinformatics and Chemical Engineering* and is an editorial member of the *International of Journal of Advanced Packaging Technology*. He has published many articles and presented at many professional conferences.

REVIEWERS AND ADVISORY BOARD MEMBERS

Gennady E. Zaikov, DSc

Gennady E. Zaikov, DSc, is Head of the Polymer Division at the N. M. Emanuel Institute of Biochemical Physics, Russian Academy of Sciences, Moscow, Russia, and Professor at Moscow State Academy of Fine Chemical Technology, Russia, as well as Professor at Kazan National Research Technological University, Kazan, Russia. He is also a prolific author, researcher, and lecturer. He has received several awards for his work, including the Russian Federation Scholarship for Outstanding Scientists. He has been a member of many professional organizations and on the editorial boards of many international science journals.

A. K. Haghi, PhD

A. K. Haghi, PhD, holds a BSc in urban and environmental engineering from University of North Carolina (USA); a MSc in mechanical engineering from North Carolina A&T State University (USA); a DEA in applied mechanics, acoustics and materials from Université de Technologie de Compiègne (France); and a PhD in engineering sciences from Université de Franche-Comté (France). He is the author and editor of 65 books as well as 1000 published papers in various journals and conference proceedings. Dr. Haghi has received several grants, consulted for a number of major corporations, and is a frequent speaker to national and international audiences. Since 1983, he served as a professor at several universities. He is currently Editor-in-Chief of the *International Journal of Chemoinformatics and Chemical Engineering* and *Polymers Research Journal* and on the editorial boards of many international journals. He is a member of the Canadian Research and Development Center of Sciences and Cultures (CRDCSC), Montreal, Quebec, Canada.

CONTENTS

LIST OF CONTRIBUTORS

Arezo Afzali
University of Guilan, Rasht, Iran

Dariusz M. Bieliński
Institute of Polymer and Dye Technology, Technical University of Łódź, Poland, Institute for Engineering of Polymer Materials and Dyes, Division of Elastomers and Rubber Technology, Piastów, Poland, Tel: +4842 6313214, Fax: +4842 6362543, E-mail: dbielin@p.lodz.p

N. I. Chekunaev
Semenov Institute of Chemical Physics, Russian Academy of Sciences (RAS), U l. Kosygina, 4, Moscow, 119 991, Russia. E-mail: amkaplan@mail.ru

Jakub Czakaj
AIB Ślączka, Szpura, Dytko spółka jawna, Knurów

Jacek Grams
Institute of General and Ecological Chemistry, Technical University of Łódź, Stefanowskiego 12/16, 90-924 Łódź, Poland

A. K. Haghi
Department of Textile Engineering, University of Guilan, P.O. Box: 41635-3756, Rasht, Iran

A. M. Kaplan
Semenov Institute of Chemical Physics, Russian Academy of Sciences (RAS), U l. Kosygina, 4, Moscow, 119991, Russia. E-mail: amkaplan@mail.ru

V. V. Kasparov
Emanuel Institute of Biochemical Physics of RAS, Russian Federation, 119334, Moscow, Kosygin St. 4.; E-mail: rasum@sky.chph.ras.ru; Tel: +7 (495) 9397320

Zinaida S. Khasbulatova
Chechen State Pedagogical Institue, 33 Kievskaya str., Groznyi 364037, Chechnya, Russia, Russian Federation; E-mail: hasbulatova@list.ru

A. L. Kovarski
Emanuel Institute of Biochemical Physics of RAS, Russian Federation, 119334, Moscow, Kosygin St. 4.; E-mail: rasum@sky.chph.ras.ru; Tel: +7 (495) 9397320

Shima Maghsoodlou
University of Guilan, Rasht, Iran

Babak Noroozi
Department of Textile Engineering, University of Guilan, P.O. Box: 41635-3756, Rasht, Iran

Saeedeh Rafiei
Department of Textile Engineering, University of Guilan, P.O. Box: 41635-3756, Rasht, Iran

Maria Rajkiewicz
Institute for Engineering of Polymer Materials and Dyes Department of Elastomers and Rubber Technology in Piastów

S. D. Razumovskii
Emanuel Institute of Biochemical Physics of RAS, Russian Federation, 119334, Moscow, Kosygin St. 4.; E-mail: rasum@sky.chph.ras.ru; Tel: +7 (495) 9397320

Mariusz Siciński
Institute of Polymer and Dye Technology, Technical University of Łódź, Stefanowskiego 12/16, 90-924 Łódź, Poland

Marcin Ślączka
AIB Ślączka, Szpura, Dytko spółka jawna, Knurów

Czesław Ślusarczyk
Institute of Textile Engineering and Polymer Materials, University of Bielsko-Biała, Poland

Michał Wiatrowski
Department of Molecular Physics, Technical University of Łódź, Stefanowskiego 12/16, 90-924 Łódź, Poland

Gennady E. Zaikov
N.M. Emanuel Institute of Biochemical Physics, 4 Kosygin str., 119334 Moscow, Russia E-mail: Chembio@sky.chph.ras.ru

V. A. Zhorin
Semenov Institute of Biochemical Physics of RAS, Russian Federation, 119334, Moscow, Kosygin St. 4; E-mail: ichph@chph.ras.ru; Tel: +7 (499) 137 67 11

LIST OF ABBREVIATIONS

AC	Asphalt Concrete
ACHF	Activated Carbon Hollow Fibers
ACNF	Activated Carbon Nanofiber
ACs	Activated Carbons
AF	Attractive Force
AN	Acrylonitrile
AP	Aromatic Polyesters
APP	Asymmetric Packing Process
BET	Brunner-Emmett-Teller
BJH	Barrett–Joiner–Halenda
CNF	Carbon Nanofibers
CNTs	Carbon Nanotubes
CVD	Chemical Vapor Deposition
DBTL	Dibutyl Tin Dilaurate
DCP	Dicumyl Peroxide
DDT	Dichlorodiphenyltrichloroethane
DF	Directional Force
DMF	Dimethyl Formaldehyde
DPPG	Diphenylpycrylhydrazyl
DSC	Differential Scanning Colorimetry
ED	External force-induced Directional factor
EDLC	Electrochemical Double-Layer Capacitors
EDXS	Energy dispersive X-ray spectrometry
EELS	Electron Energy Loss Spectroscopy
EF-F	External Force-specific Functional segment
EOE	Ethylene-Octane Elastomer
EPDM	Ethylene-Propylene-Diene Rubber
ESR	Electron Spin Resonance
EV	Electric Vehicles
F-BU	Fabrication Building Unit
FeS	Iron Sulphide
FET	Field Effect Transistors

FTIR	Fourier Transform Infrared
GIXRD	Grazing Incidence X-ray Diffraction
HK	Horvath–Kawazoe
HR	Heat Radiation
HTT	Heat-Treatment Temperature
iPP	Isotactic Polypropylene
LDPE	Low Density Polyethylene
LIB	Lithium Ion Battery
LQPs	Liquid-Crystal Polyesters
MFCs	Microbial Fuel Cells
MMA	Methyl Methacrylate
MSA	Molecular Self-Assembly
MWNTs	Multiwall Carbon Nanotubes
N-CE	Nano-Communication Element
NF	Nanofiltration
N-ME	Nano-Mechanical Element
N-PE	Nano-Property Element
N-SE	Nano-Structural Element
PAN	Polyacrylonitrile
PHEV	Plug-in Hybrid Electric Vehicles
PMMA	Poly (methyl methacrylate)
PP	Polypropylene
Pt	Platinum
PVDC	Polyvinylidene Chloride
R-BU	Reactive Building Unit
RF	Repulsive Force
RHR	Rate of Heat Release
SA-BU	Self-Assembly Building Unit
SAMs	Self-Assembled Monolayers
SBR	Styrene-Butadiene Rubber
SEM	Scanning Electron Microscopy
SMPE	Sulfophenyl Methallyl Ether
SSS	Sodium P-Styrene Sulfonate
SWNTs	Single Wall Carbon Nanotubes
TEM	Transmission Electron Microscopy
TGA	thermogravimetric analysis
TOF-SIMS	Time of Flight Secondary Ion Mass Spectroscopy
TPEs	Thermoplastic Elastomers

TPE-V	Thermoplastic Vulcanisates
VA	Vinyl Acetate
VLS	Vapor-Liquid-Solid

LIST OF SYMBOLS

a	attractive segment
a(k)	distribution of wave numbers
a_s	crack radius
d	directional segment
d^3r	volume element
de	exact differential energy
dp	probability
dp_s	differential entropy for macrosystems
e	modulus of elasticity
E'	storage modulus
E"	loss modulus
i	intensity
k	wave vector
l	long period
m	mass
r	repulsive segment
S	entropy for nano/small systems
t_c	crystallization
tm	temperature of melting
v	rate of crack propagation
v_m	limiting crack velocity

Greek Symbols

Θ	half of the scattering angle
ρ	material density
τ	time interval
ν	velocity
Δh°_m	enthalpy of melting of polyethylene crystal
Δh_m	enthalpy of melting
Δk	stress intensity factor change
δq_{in}	inexact differential amounts of heat
δw_{in}	inexact differential amounts of work
Δx	spatial extension

ψ	wave function
λ	wavelength
$\epsilon_{100\%}$	tension at 100% elongation
ϵ_B	relative elongation at the tear off
ϵ_y	elongation on yield point
λ	wavelength of the x-rays
σ_M	maximum tension
σ_y	yield point
ω	frequency

PREFACE

Interdisciplinary research and development (IR&D) is research and development that involve interaction among two or more disciplines. IR&D is needed to meet the demand of problems that cannot be solved by using any single discipline.

The aim of this book is to present important aspects of applied research and evaluation methods in chemical engineering and materials science that are important in chemical technology and in the design of chemical and polymeric products.

This book gives readers a deeper understanding of physical and chemical phenomena that occur at surfaces and interfaces. Important is the link between interfacial behavior and the performance of products and chemical processes.

This book presents original experimental and theoretical results on the leading edge of applied research and evaluation methods in chemical engineering and materials science. Each chapter has been carefully selected in an attempt to present substantial research results across a broad spectrum.

This volume presents some fascinating phenomena associated with the remarkable features of high performance nanomaterials and also provides an update on applications of modern polymers.

The book helps to fill the gap between theory and practice. It explains the major concepts of new advances in high performance materials and their applications in a friendly, easy-to-understand manner.

This volume provides both a rigorous view and a more practical, understandable view of chemical compounds and chemical engineering and their applications.

This new book:

- highlights some important areas of current interest in polymer products and chemical processes;
- focuses on topics with more advanced methods;
- emphasizes precise mathematical development and actual experimental details;

- analyzes theories to formulate and prove the physicochemical principles; and
- provides an up-to-date and thorough exposition of the present state-of-the-art of complex materials.

CHAPTER 1

STRUCTURAL ASPECTS OF COMPONENTS CONSTITUTING LOW DENSITY POLYETHYLENE/ ETHYLENE-PROPYLENE-DIENE RUBBER BLENDS

DARIUSZ M. BIELIŃSKI, and CZESŁAW ŚLUSARCZYK

CONTENTS

ABSTRACT

Structural aspects of components constituting low-density polyethylene/ethylene-propylene-diene rubber (LDPE/EPDM) blends are studied in bulk and compared to the surface layer of materials. Solvation of a crystalline phase of LDPE by EPDM takes place. The effect is more significant for systems of amorphous matrix, despite a considerable part of crystalline phase in systems of sequenced EPDM matrix seems to be of less perfect organization. Structural data correlate perfectly with mechanical properties of the blends. Addition of LDPE to EPDM strengthens the material. The effect is higher for sequenced EPDM blended with LDPE of linear structure.

The surface layer exhibits generally lower degree of crystallinity in comparison to the bulk of LDPE/EPDM blends. The only exemption is the system composed of LDPE of linear structure and amorphous EPDM exhibiting comparable values of crystallinity. Micro indentation data present the negative surface gradient of hardness. Sequenced elastomer matrix always produces significantly lower degree of crystallinity, no matter LDPE structure, whereas systems of amorphous EPDM matrix follow the same trend only when branched polyethylene of lower crystallinity is added. Values of long period for the blends are significantly higher than that of their components, what suggest some part of ethylene sequences from elastomer phase to take part in recrystallization. LDPE of linear structure facilitates the phenomenon, especially if takes in amorphous EPDM matrix. Branched LDPE recrystallizes to the same lamellar thickness, no matter the structure of elastomer matrix.

1.1 INTRODUCTION

Polyolefine blends are group of versatile materials, which properties can be tailored to specific applications already at the stage of compounding and further processing. Our previous papers on elastomer/plastomer blends were devoted to phenomenon of co crystallization in isotactic polypropylene/ethylene-propylene-diene rubber (iPP/EPDM) [1] or surface segregation in low-density polyethylene/ethylene-propylene-diene rubber (LDPE/EPDM) [2, 3] systems. Composition and structure of the materials were related to their properties. Recently, we have described the influence of molecular weight and structure of components on surface segregation of LDPE in blends with EPDM, and morphology of the surface layer being formed [4].

This paper completes the last one with structural data, calculated from X-rays diffraction spectra, collected for bulk and for the surface layer of LDPE/EPDM blends. We have focused on comparison between mechanical properties, relating them to the degree of crystallinity and a crystalline phase being formed in bulk and in the surface layer of the systems.

1.2 EXPERIMENTAL PART

1.2.1 MATERIALS

The polymers used in this study are listed in Table 1.1, together with their physical characteristics.

TABLE 1.1   Physical Characteristics of the Polymers Studied

Polymer	Density [g/cm³]	Solubility parameter [J^{0.5}/m^{1.5}]ª	Degree of branching FTIR[b]	Melting temperature [°C]	Molecular weight, M_w	Dispersity index M_w/M_n
EPDM1	0.86	15.9×10⁻³	–	44	–	–
EPDM2	0.86		–	–	–	–
LDPE1	0.930	15.4×10⁻³	3.8	112	15,000	2.32
LDPE2	0.906		6.0	90	35,000	2.56

EPDM1, EPDM2 – ethylene-propylene-diene rubber: Buna EPG-6470, G-3440 (Bayer Germany; monomer composition by weight: 71% and 48% of ethylene, 17% and 40% of propylene, 1.2% of ethylidene-norbornene, respectively).
LDPE1, LDPE2 – low-density polyethylenes (Aldrich Chemicals, UK: cat. no. 42,778–0 and 42,779–9, respectively).
ªTaken from Ref. [5].
[b]Determined from infrared spectra according to Ref. [6].

LDPE in the amount of 15phr was blended with ethylene-propylene-diene terpolymer. The method of blend preparation, at the temperature of 145 °C, that is, well above melting point of the crystalline phase of polyethylene, was described in Ref. [2]. To crosslink elastomer matrix 0.6phr of dicumyl peroxide (DCP) was admixed to the system during the second stage at 40 °C. The systems were designed in a way enabling studying the influence of molecular structure of the rubber matrix or molecular weight, crystallinity/branching of the plastomer on mechanical properties of LDPE/EPDM blends, both: in bulk and exhibited by the surface layer. Structural branching of the polyethylenes studied was simulated from their ^{13}C NMR spectra applying the Cherwell Scientific (UK) NMR software. LDPE1 contains short branches, statistically every 80 carbons in the backbone. It was found that every fourth branch is longer, being constituted of 6–8 carbon atoms. LDPE2, characterized by higher degree of branching, however contains short branch of 2–4 carbon atoms, but placed ca. every 15-backbone carbons. Samples were steel mold vulcanized in an electrically heated press at 160 °C, during time $t_{0.9}$ determined rheometrically, according to ISO 3417.

1.2.2 TECHNIQUES

1.2.2.1 X-RAYS DIFFRACTION

WAXS and SAXS measurements were carried out applying the same equipment and procedures described in Ref. [1].

1.2.2.1.1 WAXS

Investigations were performed in the scattering angle range 5–40° with a step of 0.1°. Each diffraction curve was corrected for polarization, Lorentz factor and incoherent scattering. Each measured profile was deconvoluted into individual crystalline peaks and an amorphous halo, following the procedure described by Hindeleh and Johnson [7]. Fitting was realized using the method proposed by Rosenbrock and Storey [8]. The degree of crystallinity was calculated as the ratio of the total area under the resolved crystalline peaks to the total area under the unresolved X-rays scattering curve.

1.2.2.1.2   SAXS

Measurements were carried out over the scattering angle $2\Theta=0.09–4.05°$ with a step of $0.01°$ or $0.02°$, in the range up to and above $1.05°$, respectively. Experimental data were smoothed and corrected for scattering and sample absorption by means of the computer software FF SAXS-5, elaborated by Vonk [9]. After background subtraction, scattering curves were corrected for collimation distortions, according to the procedure proposed by Hendricks and Schmidt [10, 11]. Values of the long period (L) were calculated from experimental data, using the Bragg equation:

$$L=\lambda/ (2 \sin\Theta) \qquad\qquad (1)$$

where: L – long period, Θ – half of the scattering angle, λ – wavelength of the X-rays.

1.2.2.1.3   GIXRD

Grazing incidence X-ray diffraction (GIXRD) experiments were carried out at room temperature using a Seifert URD-6 diffractometer equipped with a DSA 6 attachment. $CuK\alpha$ radiation was used at 30kV and 10 mA. Monochromatization of the beam was obtained by means of a nickel filter and a pulse-height analyzer. The angle of incidence was fixed to $0.5°$, so that the X-rays penetration into the sample could be kept constant during measurements. Diffraction scans were collected for 2Θ values from $2°$ to $60°$ with a step of $0.1°$.

1.2.2.2   DIFFERENTIAL SCANNING COLORIMETRY (DSC)

Enthalpies of melting were determined with a NETZSCH 204 differential scanning calorimeter (Germany), calibrated for temperature and enthalpy using an indium standard. Specimens of about 9–10 mg were frame cut from sheets of a constant thickness to eliminate possible influence of the specimen geometry on a shape of DSC peak. Experiments were carried out over the temperature range 30–160 °C. Prior to cooling down, the samples were kept for 5 min at 160 °C. Melting and crystallization were carried out with a scanning rate of 10deg/min. Temperature of melting Tm or crystallization T_c were taken as the ones corresponding to 50% of the adequate transition. The enthalpy of melting – ΔH_m or crystallization – ΔH_c were taken as areas under the melting or crystallization peak, respectively. The degree of the blend crystallinity was calculated from a ΔH_m value, according to the formula:

$$X_c = \frac{\Delta H_m}{\Delta H_m^o} \qquad (2)$$

where ΔH_m^o stands for the enthalpy of melting of polyethylene crystal: 289 J/g [12].

1.2.2.3 MECHANICAL PROPERTIES

Mechanical properties of polymer blends during elongation were determined with a Zwick 1435 instrument, according to ISO 37. Dumb-bell specimen geometry n°2 was applied.

1.2.2.4 NANOINDENTATION

Hardness and mechanical moduli of polymer blends were determined with a Micro materials Nano Test 600 apparatus (UK), applying the procedure of spherical indentation with 10% partial unloading. R=5 μm stainless steel spherical indenter probed the surface layer of material with the loading speed of dP/dt=0.2 mN/s, reaching depths up to 8.0 μm. More information on the instrumentation can be found elsewhere [13].

1.3 RESULTS AND DISCUSSION

1.3.1 SUPRAMOLECULAR STRUCTURE IN BULK

Degree of bulk crystallinity, calculated for the blends from DSC and WAXS spectra, are given in Table 1.2.

TABLE 1.2   Degree of Bulk Crystallinity of the Materials Studied

Sample	Degree of crystallinity [%]			
	WAXS exp.	WAXS calc.	DSC exp.l	DSC calc.
EPDM1	20.5	–	3.9	–
EPDM2	0	–	0	–
LDPE1	47.3	–	61.4	–
LDPE2	30.6	–	32.0	–
LDPE1/EPDM1	22.4	24.0	8.4	11.4
LDPE2/EPDM1	19.0	21.8	5.0	10.0
LDPE1/EPDM2	2.7	6.2	4.8	8.0
LDPE2/EPDM2	1.6	4.0	2.8	4.2

They differ significantly from the values calculated additively. Solvation of a crystalline phase of plastomer by elastomer matrix seems to be apparent. The effect, exceeding 50%, is significant in the case of amorphous matrix (EPDM2), whereas in the case of sequenced elastomer matrix (EPDM1) the difference between degrees of crystallinity calculated additively and determined by WAXS is less than 10%, approaching experimental error.

DSC spectra of LDPE/EPDM blends follow the WAXS analysis in the case of system with amorphous matrix (EPDM2). For the blends of sequenced elastomer matrix (EPDM1) the degree of crystallinity calculated from heat of melting data is significantly lower than determined from X-rays wide angle diffraction and points definitely on the crystalline phase swelling phenomenon to take place. In our opinion the huge difference in the degree of crystallinity of EPDM1 between calculations from DSC ($X_c = 3.9\%$) and WAXS ($X_c = 20.5\%$) spectra is responsible for the former, whereas limited resolution of calorimetry to less perfect organization of macromolecules (paracrystalline phase) seems to be a justification for the latter. Packing of macromolecules, is of high order for the plastomers studied, whereas for sequenced elastomer seems to be quite low, judging from the mentioned already huge difference on X_c of EPDM1, depending on the applied method of analysis.

1.3.2 MECHANICAL PROPERTIES

Mechanical properties of the blends are given in Table 1.3.

TABLE 1.3   Mechanical Properties of the Materials Studied

Sample	S_{100} [MPa]	S_{200} [MPa]	S_{300} [MPa]	TS [MPa]	E_b [%]
EPDM1	3.3	4.5	6.0	6.9	508
EPDM2	1.0	1.2	1.4	4.0	666
LDPE1/EPDM1	3.7	4.1	4.7	20.4	555
LDPE2/EPDM1	3.1	3.6	4.5	17.5	902
LDPE1/EPDM2	1.7	1.9	2.0	4.9	852
LDPE2/EPDM2	1.6	1.8	1.9	4.3	792

They correlate well with structural data. The higher the degree of crystallinity the higher the moduli in extension, tensile strength of material and elongation at break. Addition of LDPE to EPDM improves mechanical properties of the elastomer, what is especially visible for the sequenced matrix. Strengthening effect is more pronounced when lower molecular weight but of higher crystallinity plastomer is added.

1.3.3 SUPRAMOLECULAR STRUCTURE IN THE SURFACE LAYER

Degree of crystallinity of the surface layer of blends, calculated from WAXS spectra (low incidence angle) are given in Table 1.4.

TABLE 1.4   Degree of the Surface Layer Crystallinity of the Materials Studied

Sample	Degree of crystallinity [%]	
	WAXS exp.	WAXS calc.
EPDM1	9.6	–
EPDM2	0	–
LDPE1	47.3	–

TABLE 1.4 *(Continued)*

Sample	Degree of crystallinity [%]	
	WAXS exp.	WAXS calc.
LDPE2	32.0	–
LDPE1/EPDM1	10.5	14.5
LDPE2/EPDM1	6.1	12.3
LDPE1/EPDM2	5.6	6.2
LDPE2/EPDM2	1.3	4.0

The surface layer of both: polymer components and their blends, exhibits generally, in comparison to the bulk, lower values of the degree of crystallinity. The only exemption is the LDPE1/EPDM2 system, for which the crystallinity of the surface layer is significantly higher than in bulk and additive calculations give the value only slightly higher in comparison to the experimental one.

Comparing the WAXS data determined in bulk to the ones characterizing the surface layer of the systems studied, one can find that their relation does depend on supermolecular structure of components. Sequenced elastomer matrix always produces significantly lower than in bulk degree of crystallinity, no matter the structure of plastomer, whereas the same is followed by amorphous elastomer matrix only when branched polyethylene (LDPE2) of lower crystallinity is added. Amorphous EPDM matrix facilitates crystallization of low molecular weight polyethylene of higher crystallinity (LDPE1) on the surface.

Values of long period (L) for the blends are significantly higher than that of their components Table 1.5.

TABLE 1.5 Values of the Long Period (L) of the Materials Studied (SAXS)

Sample	L [nm]
EPDM1	14.3
EPDM2	0
LDPE1	14.5
LDPE2	10.6

TABLE 1.5 *(Continued)*

Sample	L [nm]
LDPE1/EPDM1	17.8
LDPE2/EPDM1	14.2
LDPE1/EPDM2	24.4
LDPE2/EPDM2	14.2

It suggests some part of ethylene sequences from the elastomer phase to take part in recrystallization. Polyethylene of higher linearity (LDPE1) facilitates the phenomenon, which takes place to the higher extent in the amorphous elastomer matrix (EPDM2). Plastomer of higher branching (LDPE2) recrystallizes to the same lamellar thickness, no matter the structure of the elastomer matrix.

1.3.4 NANOINDENTATION

Hardness profiles of the surface layer of blends studied are presented in Figs. 1.1 and 1.2.

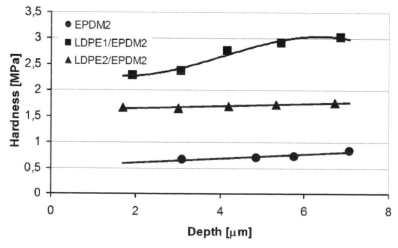

FIGURE 1.1   Hardness profile of LDPE/EPDM1 blends.

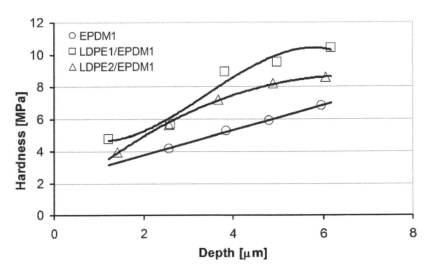

FIGURE 1.2   Hardness profile of LDPE/EPDM2 blends.

They confirm structural data Blends of sequenced elastomer matrix (EPDM1) exhibit considerably higher hardness in comparison to the systems of amorphous one (EPDM2). It concerns both: bulk as well as the surface layer of materials. Improvement of mechanical properties of EPDM by blending with LDPE is easier to achieve when amorphous elastomer is to be modified. Linear polyethylene (LDPE1) is better than more branched one (LDPE2) in terms of reaching higher hardness. The difference is especially pronounced in bulk, gradually diminishing towards the surface of materials.

1.4   CONCLUSIONS

1. EPDM matrix solvates crystalline phase of LDPE. The effect is significant for blends of amorphous elastomer matrix. In the case of sequenced EPDM matrix a part of crystalline phase of LDPE recrystallizes in less perfect form, which is not detectable by DSC.
2. Improvement of mechanical properties of EPDM by blending with LDPE is higher for the elastomer of sequenced structure, confirming structural data. The best mechanical properties, obtained when linear plastomer is admixed to amorphous elastomer, stays

in agreement with the highest degree of crystallinity of the systems studied.

3. Degree of crystallinity of polymer components and their blends are higher in bulk than in the surface layer. Blends containing sequenced EPDM matrix exhibit the most significant difference, no matter molecular structure of LDPE, whereas the systems containing amorphous elastomer matrix facilitates recrystallization of linear polyethylene on the surface.

4. Lamellas of crystalline phase of the surface layer of poly olefin blends studied are thicker than present in the surface layer of their components, what suggests co crystallization of ethylene monomer unit from EPDM. Linear LDPE facilitates the phenomenon, especially when takes place in amorphous elastomer matrix. Branched plastomer re crystallizes to the same lamellar thickness, no matter the structure of elastomer matrix.

5. Micro indentation data reveals hardness profile of LDPE/EPDM blends, staying in agreement with structural data for the surface layer of systems studied.

KEYWORDS

- **Mechanical properties**
- **Polyolefine blends**
- **Structure**
- **Surface layer**

REFERENCES

1. Bieliński, D., Ślusarski, L., Włochowicz, A., Ślusarczyk, Cz., & Douillard, A (1997). *Polimer Int. 44*, 161.
2. Bieliński, D., Ślusarski, L., Włochowicz, A., & Douillard, A. (1997). *Composite Interf., 5*, 155.
3. Bieliński, D., Włochowicz, A., Dryzek, J., & Ślusarczyk, Cz. (2001). *Composite Interf., 8*, 1.
4. Bieliński, D., & Kaczmarek, Ł. (2006). *J. Appl. Polym. Sci, 100*, 625.
5. Barton, A. F. M. (1981). *Handbook of Solubility Parameters and Other Cohesion Parameters*, CRC Press, Boca Raton, FL.

6. Bojarski, J., & Lindeman, J. (1963). *Polyethylene, 109,* WNT, Warsaw.
7. Hindeleh, A. M., & Johnson, J. (1971). *J. Phys. D: Appl. Phys., 4,* 259.
8. Rosenbrock, H. H., & Storey, C. (1966). *Computational Techniques for Chemical Engineers,* Pergamon Press, New York.
9. Vonk, C. G. (1970). *J. Appl. Crystal., 8,* 340.
10. Hendricks, R. W., & Schmidt, P. W. (1967). *Acta Phys. Austriaca, 26,* 97.
11. Hendricks, R. W., & Schmidt, P. W. (1970). *Acta Phys. Austriaca, 37,* 20.
12. Brandrup, J., & Immergut, E. H. (1989). *Polymer Handbook 3rd Ed.*, Ch. 5, John Wiley & Sons, London-New York.
13. www.micromaterials.com

CHAPTER 2

COMPARISON OF POLYMER MATERIALS CONTAINING SULFUR TO CONVENTIONAL RUBBER VULCANIZATES IN TERMS OF THEIR ABILITY TO THE SURFACE MODIFICATION OF IRON

DARIUSZ M. BIELIŃSKI, MARIUSZ SICIŃSKI, JACEK GRAMS, and MICHAŁ WIATROWSKI

CONTENTS

ABSTRACT

The degree of modification of the surface layer of Armco iron by sulfur, produced by sliding friction of the metal sample against: ebonite, sulfur vulcanizate of styrene-butadiene rubber, polysulphone or polysulfide rubber, was studied. Time of Flight-Secondary Ion Mass Spectroscopy (TOF-SIMS) and confocal Raman microscopy techniques, both confirmed on the presence of iron sulfide (FeS) in the surface layer of metal counter face after tri biological contact with SBR or polysulfide rubber. For the friction couple iron-ebonite, the presence of FeS was confirmed only by TOF-SIMS spectra. FT-Raman analysis indicated only on some oxides and unidentified hydrocarbon fragments being present. Any sulfur containing species were not found in the surface layer of iron counter face due to friction of the metal against polysulphone. The degree of iron modification is determined by the loading of friction couple, but also depends on the way sulfur is bonded in polymer material. Possibility for modification is limited only to materials, which contain sulfur either in a form of ionic sulfide crosslinks (SBR and ebonite) or side chains (polysulfide rubber). Degradation of polymer macromolecules during friction (polysulphone and ebonite in this case under high loading) does not lead to the formation of FeS. Chemical reaction between sulfur and iron takes place only in the case of ionic products of polymer destruction containing sulfur.

2.1 INTRODUCTION

Interest towards chemical reactions accompanying friction has been growing in the last few years. This is reflected by significant progress in very important area of tribiology, called tribochemistry [1]. One of its priorities are studies on chemical reactions taking place in the surface layer of materials constituting the friction couple and their exploitation consequences, for example, concerning creation of protective layers, lowering wear, etc.

An increase of temperature in tribiological contact during friction is well known. It facilitates the phenomenon of selective transfer of polymer components, followed by their chemical reaction with the surface layer of metal counterface, in the case of rubber-metal friction couple. The modification cannot only effect composition and structure of the surface layer of polymer but metal as well [2]. Our previous studies confirmed on the possibility to modify the surface layer of iron counter face by sliding friction against sulfur vulcanizates of styrene-butadiene rubber (SBR) [3, 4]. Extend of modification is related to the kind of dominated sulfur crosslinks. An increase of temperature accompanying friction facilities breaking of cross links present in vulcanizate, especially polysulfide ones [5]. The highest degree of modification was detected for Armco iron specimen working in tribiological contact with rubber crosslinked by an effective sulfur system of short: mono- and di- to long polysulfide crosslinks ratio equal to 0.55. Polysulfide cross-links characterize themselves by the lowest energy from the range of cross-links created during conventional sulfur vulcanization (-C-C-, -C-S-C-, -C-S$_2$-C-, -C-S$_n$-C-; where $n \geq 3$) [6]. So, their breaking as first is the most probable. As a result, the release of sulfur ions, representing high chemical reactivity to iron, takes place. FeS layer of 100–150 nm thickness was detected on iron specimen subjected to friction against sulfur vulcanizates of SBR [3]. It lubricates efficiently the surface of metal, reducing the coefficient of friction [7]. As a compound of low shearing resistance, FeS is easily spreaded in the friction zone, adheres to metal counter face, penetrating its micro roughness. Even very thin layer of FeS showed to be effective due to high adhesion to iron. Metal oxides (mainly Fe$_3$O$_4$) being created simultaneously on the metal surface, act synergistically with FeS, making it wear resistance significantly increased [8]. This paper is to compare other polymer materials containing sulfur to conventional rubber vulcanizates in terms of their ability to the surface modification of iron. The polymers studied vary from sulfur vulcanizates either according to crosslink density (ebonite) or the kind of sulfur incorporation in macromolecules (polysulphone rigid material and polysulfide rubber elastomer).

2.2 EXPERIMENTAL PART

2.2.1 MATERIALS

Surface polished specimen made of Armco iron were subjected to extensive friction against:
- polysulphone PSU 1000 (Quadrant PP, Belgium),
- crosslinked polysulfide rubber LP-23 (Toray, Japan),
- ebonite based on natural rubber [9], or

• carbon black filled sulfur vulcanizate of SBR Ker 1500 (Z. Chem. Dwory, Poland).

Composition of the materials studied is given in Table 2.1.

TABLE 2.1   Composition of the Polymer Materials Studied

Material Components	Polysulfide rubber	Ebonite	Conventional rubber (SBR)	Polysulphone
Styrene-butadiene rubber, Ker 1500			100	
Natural rubber, RSS II		100		
Polysulfide rubber, LP-23	100			
Poly(sulphone), PSU 1000				100
Stearic acid			1	
Zinc oxide, ZnO			3	
HAF carbon black, Corax N 326			50	
Ebonite powder		50		
Linseed oil		2		
Isostearic acid	0.20			
Manganese dioxide, MnO_2	10			
Tetramethylthiuram disulfide, TMTD	0.50			
N-third buthyl-di-benzothiazolilosulphenamide, TBBS			2.20	
Zinc dithiocarbamate, Vulkacit		1		
Sulfur, S_8		42	0.80	

Rubber mixes were prepared with a David Bridge (UK) roller mixer. Specimen for further examinations was vulcanized in a steel mold at 160 °C, during time $\tau_{0.9}$, determined rheometrically with a WG 05 instrument (Metalchem, Poland), according to ISO 3417. Liquid polysulfide rubber was cured at room temperature by means of chemical initiator, activated by MnO_2. Polysulphone specimens were prepared by cutting off from a rod.

Modification of Armco iron counterface was performed by rubbing of polymer materials studied against metal specimen. The process was realized with a T-05 tribometer (IteE-PIB, Poland).

2.2.2 TECHNIQUES

2.2.2.1 TIME OF FLIGHT-SECONDARY ION MASS SPECTROSCOPY (TOF-SIMS)

Studies were carried out by means of an ION-TOF SIMS IV instrument (Germany), operating with a pulse ^{69}Ga ion gun of beam energy 25kV. Primary ion dose was any time kept below 3×10^{11} cm^{-2} (static mode). Negative and positive ion spectra of iron specimen were collected in the range of m/z 1–800, before and after friction against polymer materials. Analysis was narrowed to the range of m/z<35, which showed to be the most relevant in terms of sulfur modification. The most informative signals can be subscribed to: H$^-$ (m/z = 1), C$^-$ (m/z = 12), CH$^-$ (m/z = 13), O$^-$ (m/z = 16), OH$^-$ (m/z = 17), C$_2^-$ (m/z = 24), S$^-$ (m/z = 32) and SH$^-$ (m/z = 33). Any time counts of S$^-$ and SH$^-$ were normalized to the total number of counts present in the spectrum.

2.2.2.2 RAMAN SPECTROSCOPY

Studies were carried out by means of a Jobin-Yvon T64000 (France) instrument, operating with a laser of 514.5 nm line and power of 50–100W. The spectrometer was coupled with a BX40 Olympus confocal microscope, operating with Olympus LMPlanFI 50* (NA=0.50) or LMPlan 50* (NA=0.75) objectives. The surface of iron specimen was examined with acquisition time of 240–540 s, at least in two distant places, before and after friction. The Internet RASMIN database [10] was used for material identification. Characteristic absorption bands for FeS are present at wavelengths of 270 and 520 cm^{-1} (Fig. 2.1).

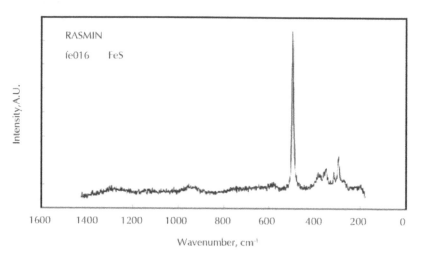

FIGURE 2.1   FT-Raman spectrum of iron sulfide [10].

2.2.2.3   TRIBOLOGICAL CHARACTERISTICS

Tribological characteristics of the materials studied were determined with a T-05 (ITeE-PIB, Poland) tribometer, operating with a block-on-ring friction couple. Ring made of polymer material was rotating over a still block made of Armco iron. The instrument worked together with a SPIDER 8 Hottinger Messtechnik (Germany) electronic system for data acquisition. The way for data analysis has been described in Ref. [11]. Polymer rings of diameter 35 mm, rotating with a speed of 60rpm were loaded within the range of 5–100N, during 60–120 min.

2.3   RESULTS AND DISCUSSION

In order to explain the influence of the way sulfur is bonding in polymer materials on modification of the surface layer of iron, the metal counter-face was subjected to sliding friction against:

- polysulphone (load of 20N/ time 2 h),
- ebonite (load either 20 or 100N/ time 2 h),
- SBR vulcanizate (load 20N/ time 2 h), and
- polysulfide rubber (load 5N/ time 1 h).

The load and time of friction in the last case have to be decreased due to low mechanical strength of polysulfide rubber.

From the specific spectra of secondary ions (Fig. 2.2) and comparison between normalized counts for particular cases (Fig. 2.3) it follows, that the highest amount of sulfur, in a form of SH^- ions, was transferred to the surface layer of iron counter face by ebonite. In the case of polysulphone, due to strong sulfur bonding to macromolecular backbone (Fig. 2.4) and different from other polymers studied mechanisms of mechanodegradation, the expected effect of sulfur transfer is practically absent. The amount of iron sulfide, created in the surface layer of iron counter face depends on reactivity of sulfur containing polymer fragment being released during friction and their concentration in the friction zone. From possible substrates, involved in the creation of FeS, the highest affinity to iron exhibit polysulfide cross links and ionic products of their destruction, released from some polymer materials subjected to intensive friction against the metal counter face. They can be produced only in the case of SBR vulcanizate and ebonite, what can be explained by their chemical structure. One should pay attention to different load being applied for the polymer materials studied. In the case of unfilled polysulfide rubber, the time of friction has additionally to be limited due to low mechanical strength of the material. However, an example of ebonite, demonstrates that an increase of loading not necessarily has to lead to higher extent of modification of iron counter face during friction – (Figs. 2.2 and 2.3).

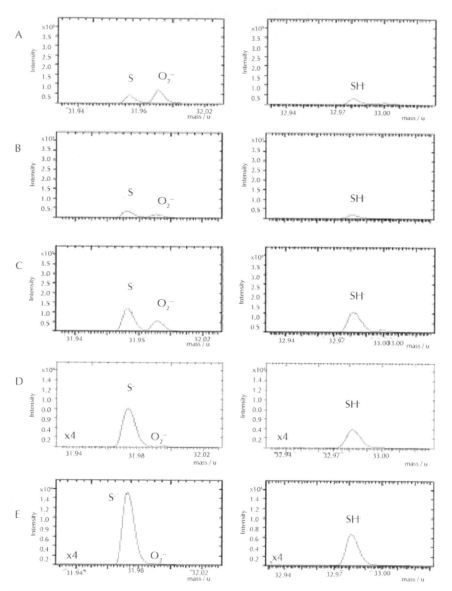

FIGURE 2.2 Specific TOF-SIMS spectra collected from the surface layer of iron Armco specimen, subjected to friction against various polymer materials studied: A – virgin, B – polysulphone, C – SBR vulcanizate, D – ebonite/100 N, E – ebonite/20N.

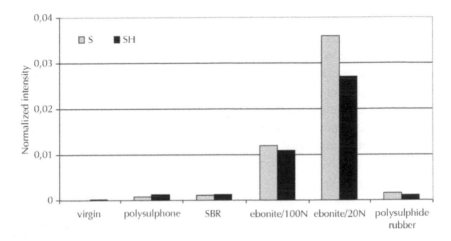

FIGURE 2.3 Normalized TOF-SIMS spectra count of S⁻ and SH⁻ ions for the surface layer of Armco iron specimen subjected to friction against various polymer materials studied.

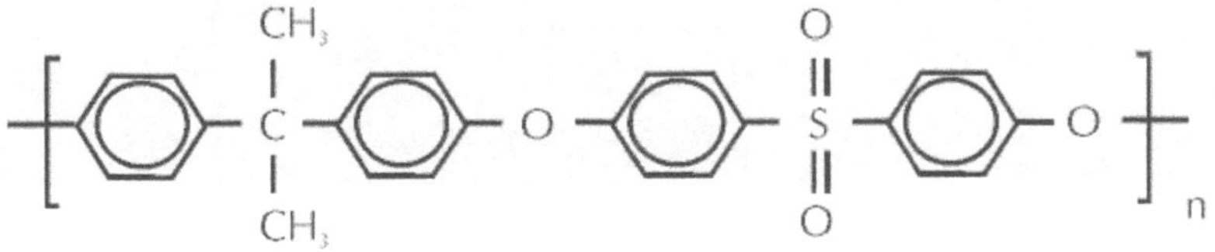

FIGURE 2.4 Chemical structure of polysulphone.

It can be the result of prevailing, under extreme friction conditions, radical degradation of macromolecules, accompanied by intensive oxidation of polymer. Distribution of all ions in the surface layer of metal is uniform. A part of sulfur ones, released from polymer materials, also oxides are formed, what facilitates antifriction properties of metal [12]. Iron sulfide present can further be oxidized during friction to sulphones, which are even better lubricating agents. However, under too high loading conditions ionic mechanism of crosslink breaking is not able to show up, losing to macromolecular degradation, what explains lower efficiency of the modification of iron with sulfur. In order to confirm TOF-SIMS data on FeS presence in the surface layer of Armco iron subjected to friction against various polymer materials (Fig. 2.5), complementary studies with Raman spectroscopy were carried out. Comparing collected spectra (Fig.

2.6) to the standard FeS spectrum from the database (Fig. 2.1), only the spectra of iron surface after friction against SBR vulcanizate and polysulfide rubber indicate on possible sulfur modification. Analysis of the surface of Armco iron specimen subjected to friction against polysulphone or ebonite did not bring unique results. FT-Raman spectra contain signals the most likely coming from degraded fragments of macromolecules or unidentified compounds containing carbon, oxygen and hydrogen. For the above two cases any absorption peak in the region characteristic for the FeS could not be assigned.

Field of view: 101.0 x 101.0 μm²

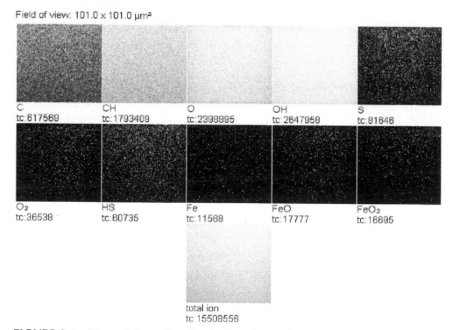

FIGURE 2.5   Maps of the surface distribution of some ions for Armco iron subjected to extensive friction against SBR vulcanizate (brighter color indicates on higher intensity of ion present).

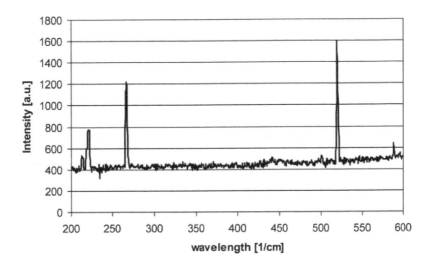

A – polysulphide rubber

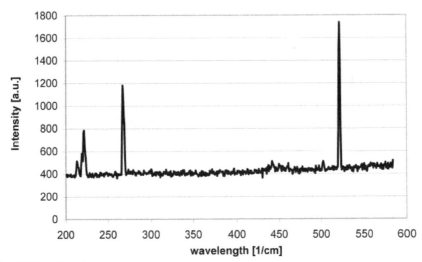

B – SBR vulcanizate

FIGURE 2.6   *(Continued)*

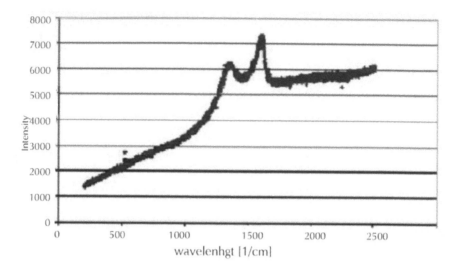

C – ebonite/100 N

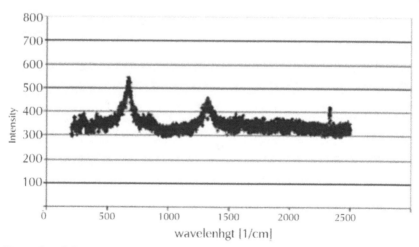

D – polysulphone

FIGURE 2.6   FT-Raman spectra of the surface layer of Armco iron specimen subjected to friction against various polymer materials studied.

Tribological characteristics of polymer materials sliding against Armco iron are demonstrated in Fig. 2.7. Friction force and energy curves vary from material to material due to their different chemical structure and related mechanical properties. From tribiological point of view, the most efficient modification has to be subscribed to ebonite-iron friction couple. In this case median of the friction force and discrete levels of energy exhibit the most stable courses among all the friction couples studied. The second run, repeated for new ebonite roll after 2hrs of previous frictional modification of Armco iron, resulted in above 10% reduction of the coefficient of friction (Fig. 2.7B). The first run for SBR vulcanizate (Fig. 2.7C) exhibits "classical" course with characteristic maximum of the coefficient of friction, which appears already after some minutes from the start of experiment. For the first 100 min value of the friction force gradually decreased from 38 down to 26N, eventually stabilizing at this level. In the second run, the friction force comes back to the initial value, but right after beginning immediately goes down to the final value after the first run. It means that in the case of SBR vulcanizates, the modification of metal counter face is the most important for the beginning of friction. Tribiological characteristics determined for polysulfide rubber (Figs. 2.7E and 2.7F) are not so stable as for ebonite or SBR vulcanizates, probably because of poor mechanical properties of polysulfide rubber. Nevertheless, the modification of the surface layer of Armco iron, confirmed by TOF-SIMS and Raman spectroscopy, is also reflected by tribiological data. In the first cycle, the friction force is maintained at the level of 17–18N during the first 25 min, and suddenly goes down to 10N, which level is kept constant till the end of experiment. The drop is reflected by significant increase of the energy component responsible for high-energy vibrations (200–600 Hz). The vibrations of such energy are not present in tribological characteristics of elastomers [11]. Similarly to SBR vulcanizate, the second run for polysulfide rubber starts from higher value of the friction force, which quickly goes down and stabilizes itself at the final level of the first run. In the case of polysulphone (Figs. 2.7G and 2.7H) any tri biological effects, able to be subscribed to the surface modification of iron counter face, have not been observed. During the first run, value of the friction force is increasing, eventually reaching stabilization at the level of 9N, shortly before the end of experiment. The second run starts from the friction force value of 5N, which gradually increases up to the final level of the first run. At this moment, this requires about 40 min from start, its course become very unstable, probably because of intensive wear of iron counter face influencing experimental data.

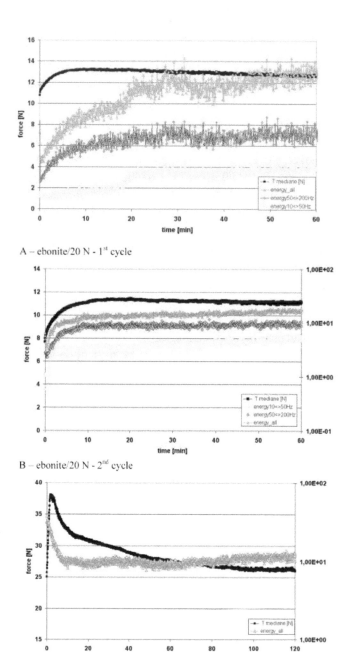

A – ebonite/20 N - 1st cycle

B – ebonite/20 N - 2nd cycle

C – SBR vulcanizate - 1st cycle

FIGURE 2.7   *(Continued)*

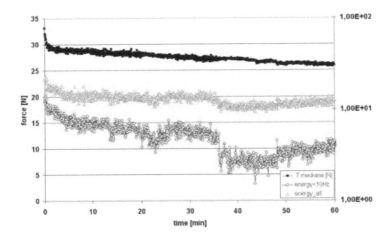

D – SBR vulcanizate - 2nd cycle

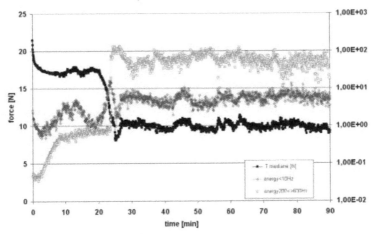

E – polysulphide rubber - 1st cycle

FIGURE 2.7   *(Continued)*

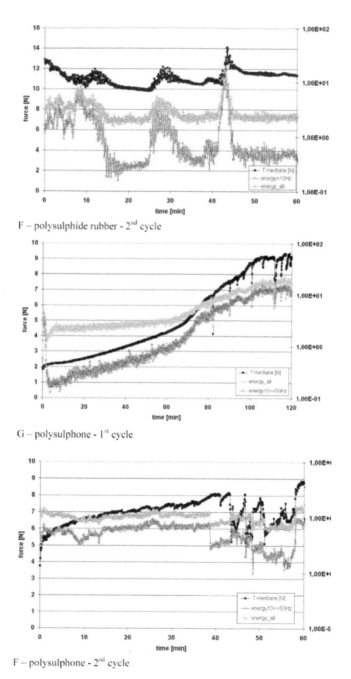

FIGURE 2.7   Tribological characteristics of Armco iron-polymer material friction couple.

2.4   CONCLUSIONS

1. Extensive friction against polymer materials containing sulfur can result in surface modification of Armco iron counter face with FeS. Extent of the modification depends on the way of sulfur bonding to macromolecules.

2. The results obtained point of higher efficiency of modification, when sulfur is present in ionic cross-links of polymer material-SBR vulcanizate and ebonite, contrary to constituting its macromole-cules-polysulphone. Degradation of the latter is of radical character and its products immediately react with atmospherics oxygen. Only ionic species, produced by breaking of sulphidic cross-links, are able to react with iron.

3. The extent of modification is straightly related to the amount of sulfur being present in cross-links. The highest amount of sulfur in the case of ebonite results in the highest degree of modification of the surface layer of Armco iron after tri biological contact. Application of cured polysulfide rubber is less effective due to the lack of sulfide cross links in structure of the material. Additionally, its poor mechanical properties are responsible for a transfer of low molecular weight products of wear onto the metal counterface, whereas "strong" polysulphone makes Armco iron sample worn off.

4. TOF-SIMS data for Armco iron point on the highest degree of sulfur modification being the result of friction against ebonite, SBR vulcanizate and cured polysulfide rubber. The spectra represent the highest amount of species containing sulfur.

5. Tribological characteristics confirm the influence of sulfur modification of metal counter face on lowering friction for the metal-polymer couples studied. In the case of ebonite, the coefficient of friction reduced significantly for the whole experimental run, whereas application of SBR vulcanizate or polysulfide rubber was effective only during the first period of experimental cycles. Any improvement of tribiological characteristics was not assigned for polysulphone. The polymer was observed to worn the surface of iron counterface, what resulted in increase of the coefficient of friction in this case.

KEYWORDS

- **Iron**
- **Polymers**
- **Sliding friction**
- **Surface modification**

REFERENCES

1. Płaza, S. (1997). *Physics and Chemistry of Tribological Processes,* University of Łódź Press, Łódź.
2. Rymuza, Z. (1986). *Tribology of Sliding Polymers,* WNT, Warsaw.
3. Bieliński, D. M., Grams, J., Paryjczak, T., & Wiatrowski, M. (2006). *Tribological Modification of Metal Counterface by Rubber,* Tribological Letters, *24,* 115–118.
4. Bieliński, D. M., Siciński, M., Grams, J., & Wiatrowski, M. (2007). Influence of the crosslink structure in rubber on the degree of modification of the surface layer of iron in elastomer-metal friction pair, *Tribological, 212,* 55–64.
5. Boochathum, P., & Prajudtake, W. (2001). Vulcanization of cis- and transpolyisoprene and their blends: cure characteristics and crosslink distribution, Eur. *Polym. J. 37,* 417–427.
6. Morrison, N. J., & Porter, M. (1983). Temperature effects on structure and properties during vulcanization and service of sulfur-crosslinked rubbers, Plast. *Rubber Proc. Appl., 3,* 295–304.
7. Grossiord, C., Martin, J. M., Le Mogne, Th., & Palermo, Th. (1998). In situ MoS formation and selective transfer from MoDPT films, *Surf. Coat Technol. 108, 109,* 352–359.
8. Wang, H., Xu, B., Liu, J., & Zhuang, D. (2005). Investigation on friction and wear behaviors of FeS films on L6 steel surface, *Appl. Surf. Sci. 252,* 1084–1091.
9. Gaczyński, R., ed. (1981). *Rubber Handbook for Engineers and Technicians,* WNT, Warsaw Tab IV-*35,* 323.
10. http://www.aist.go.jp/RIOBD/rasmin/E-index.htm
11. Głąb, P., Bieliński, D. M., & Maciejewska, K. (2004). An Attempt to Analysis of Stick-Slip Phenomenon for Elastomers, *Tribological, 197,* 43–50.
12. Wang, H., Xu, B., Liu, J., & Zhuang, D. (2005). Characterization and tribological properties of plasma sprayed FeS solid lubrication coatings, *Mater. Characterization, 55,* 43–49.

CHAPTER 3

A RESEARCH NOTE ON THE EFFECTS OF ELASTOMER PARTICLES SIZE IN THE ASPHALT CONCRETE

A. M. KAPLAN and N. I. CHEKUNAEV

CONTENTS

ABSTRACT

It is noted that the durability of asphalt concrete pavements is determined by the time of the trunk cracks formation in the polymer-containing composites in the modified by elastomers (e.g., by rubber) bitumenous binder of asphalt. Developed by the authors [1] previously the theory of the cracks propagation in heterosystems has allowed to investigate the problem of the cracks propagation in the rubber-bitumen composite. This investigations show that most effectively to prevent the trunk cracks formation in asphalt concrete can ultrafine rubber particles (150–750 nm) in a bitumenous binder of asphalt.

3.1 INTRODUCTION

The constant interest in elucidation of new possibilities to increase of the asphalt concrete pavements durability continues to persist for several decades. Such pavements are composites consisting of gravel, sand and the polymer-containing composite (bitumen) in rationally chosen ratios. According to modern notions, it is assumed that the durability of asphalt concrete pavements is determined by the appearance of numerous major trunk cracks in the pavement. This conclusion is confirmed, in particular, in numerous reports (over 130), presented at a special conference devoted exclusively to the problem of cracking in pavements [2]. It was shown earlier and in reports of the conference that a certain increase in the durability of pavements can achieve by introducing into the bitumenous binder of elastomers (synthetic rubbers or crushed technical rubber). However, very important question about a quantitative determination of the optimum size of elastomers particles introduced into the bitumen for providing of maximum durability of pavements, until recently, remained open. The theoretical solution of above problem is presented in the next section.

3.2 DEVELOPMENT OF CRACKS IN CYCLICALLY LOADED POLYMER-CONTAINING COMPOSITES

The main destruction mechanism of solids is the nucleation and growth of cracks. One of the most effective ways to improve the performance properties of the material is the introduction into bitumen of elastic modifier such as SBS or rubber in an amount of 1–10%. Asphalt concrete (AC) is a combination of different mineral components mixed with bitumen. Bitumen is a small portion of asphalt material, but plays a very important role in binding together the material, wherein the cracks are generally propagating in a bitumenous component.

In this paper the negative Celsius temperatures were assumed, when the material can be regarded as elastic solid with sufficient accuracy. In present paper were used classical notions of Griffiths and Orowan. These notions, suitable for description of the cracks behavior in homogeneous materials, were modified for description of crack development in heterogeneous materials. In heterogeneous materials, as AC, the embryonic voids with the characteristic sizes a_{ini} 20–50 nm with a concentration c_0 $=10^{17}–10^{21}$ M^{-3} [3], however, their effect on the strength and durability of the materials can be decisive.

The mechanical impact on heterogeneous materials leads existence of areas with high local stresses $\sigma_{loc} = \beta_1 \cdot \sigma_0 \approx (3 \div 10)\sigma_0$ [4], where β_1 is the overstress factor. Areas with the most elevated local stress, comparably with the average stress $\sigma_0 \sim 1.5$ MPa are located in asphalt concrete near the acuminated borders of mineral component particles in narrow inter layers of binder between the mineral particles of the asphalt. Additional stress concentrators are available in asphalt concrete macro cracks, original and also the newly formed under intense mechanical stresses.

The local stresses σ_{loc} in areas of hetero-structures spaced at distance r from a crack's tip exceed external stress in such structures by the factor $\beta_2 = \ddot{O} (a/r)$. The estimation gives $\beta_2 \sim 50$ for the region with the size of 5 microns spaced apart from the tip of the macro crack with length $a=1$ cm. Evaluation of the resultant stress near the crack's tip in the vicinity the border of mineral particles gives the value $\sigma_{loc} = \beta_2 \cdot \beta_1 \cdot \sigma_0 > 200$ MPa. These high local stresses up to 300 MPa, as noted in [5], were found also in the polymeric material. It should be noted that for larger local stress smaller spatial scale is typical. Regarding ultrahigh local stress of a few hundred MPa, the size of the corresponding area is between one and several micrometers. It is in these areas the nucleation and growth of cracks occurs, with their transformation into supercritical locally.

Rubber modifier is characterized by a slow rate of fatigue growth under external cyclic loads as well as high energy of new surface γ_R, significantly exceeding (by tens or hundreds times) of bitumen surface energy γ_B.

Therefore, the crack critical size in the rubber is significantly larger than the critical size in the bitumen $a_{cr}^{(R)} \gg a_{cr}^{(B)}$.

Important role in increasing the strength of polymers play the "crazes" formed near modifier particles at adding modifier. (See, for example, [6]). Energy expenses for the micro cracks formation allow fast enough to reduce the level of tension of the polymer sample and there by hinder its destruction. Energy spending for the micro cracks formation let to reduce fast enough the level of tension of the polymer sample and thereby hinder its destruction. Also, the presence of crazes in the vicinity of the modifier particles leads to the fact that more amount of energy is needed to create a new surface. The principle of retardation and stopping of fast propagating cracks in mechanically loaded AC pavement by particles of elastic modifier is demonstrated in Fig. 3.1.

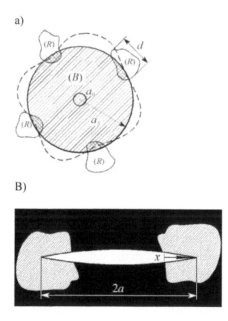

FIGURE 3.1 Schematic diagram of breaking and stopping of the disc-shaped crack at meeting with elastic modifier particles in bitumen. a) The cross section of modified bitumen sample by the plane coinciding with plane of the disk-shaped crack (shaded). Additional explanations see the text. b) The cross section of the same sample by the plane of perpendicular to the crack plane of (light). Crack surrounded with bitumen (shown in black) and partially penetrates into elastic modifier particles (shaded).

Initial microvoids are completely different shapes. Some portions α of them have a shape, which can be approximately regarded as submicro-cracks with the size a_{ini}. For some time τ they grow up to sufficiently small critical size $a_{cr}^{(B)}$ in the bitumen matrix. These cracks are disc-shaped, so the size of the crack can be characterized by its radius a. After exceeding the critical size rapid cracks growth begins in the asphalt until it meets an obstacle in the form of a particle modifier. And, as shown in Fig. 3.1, losing energy crack is stopped by modifier particles. Fatigue crack growth continues, albeit relatively slow. Cracks grow relatively fast in the bitumen and much slower when penetrating into the rubber. Cutting of elastic modifier particles by the trapped crack due to fatigue growth requires a very long time. Critical size of the disk-shaped crack is determined from the equation [7]:

$$a_{cr} = 1.6E\gamma/\sigma^2 \tag{1}$$

where E is modulus of elasticity of the material. We estimate the value of the critical radius in the various components. Values of E_B and γ_B depend on the bitumen grade and the ambient temperature. To estimate we took intermediate values $E_B = 1\,GPa$ and $\gamma_B = 10\ J/m^2$ [8], $E_R = 100\ MPa$ and $\gamma_R = 20\ kJ/m^2$ [9]. Then, when $\sigma_{loc} = 250\ MPa$ the critical radius of the disc-shaped cracks in the bitumen is equal to $a_{cr}^{(B)} = 250$ nm and in rubber $a_{cr}^{(R)} = 50$ microns. Grown to a size greater than $a_{cr}^{(B)}$, the crack can quickly accelerate and turn into a destructive trunk crack in the pure bitumen. To prevent such crack development the modifier rubber particles are incorporated in the bitumen. Let us give here briefly the theory of crack's stopping in solids by modifier particles [10]. The total energy of a disk-shaped crack of radius a:

$$a^3\left[v^2/v_m^2 - \left(1 + a_{cr}^{(B)}/(2a)\right)\left(1 - a_{cr}^{(B)}/a\right)^2\right] + 3a_{cr}^{(R)}A/(2\pi) = 0, \quad v_m^2 = cE/(k\rho) \tag{2}$$

where ρ is the material density, v is the rate of crack propagation, v_m is the limiting crack velocity, A is a new generated area at crack penetration into the rubber component. Critical crack size $a_{cr}^{(B)}$ in the bitumen component is determined from the condition $dE_{tot}(a, v=0)/da = 0$. When crack is crossing inclusions, significant role plays the last term on the left side in Eq. (2). Approximately, in accordance with Fig. 3.1, one can take $A = z\pi xd/6$, where x – an average path passed by the crack in the rubber inclusion, z 4 is the coordination number, approximately equal to the minimum number of inclusions required to stop crack. Coefficient of d is chosen so that at the full dissecting of inclusion particle ($x = d$) the formed area was equal to the average area of particles cross-section. From the Eq. (2) we can find a mean path x_S, passed the crack in the modifier particle before crack stop. At the condition $a_{cr}^{(B)} \ll a_S$, solving the equation (2), we obtain:

$$x_S = 4a_S^3 / \left(za_{cr}^{(R)} d \right) \tag{3}$$

where a_S is crack radius at which it is stopped. To stop the crack the inequality $x_S < d$ is necessary and we have the inequality:

$$d^2 > 4a_{stop}^3 / \left(za_{cr}^{(R)} \right) \tag{4}$$

The crack's circular front when pinned is strongly curved. If the crack front deviation is $\delta \times a$, then the active volume at the crack passage is equal $2\pi\delta a_S^2 d$, then the number of inclusions, caught in this volume equals $2\pi\delta da_S C_m = z$, $C_m = 3r_m^{-3}/4\pi$, where r_m is the mean distance between modifier particles, C_m is their concentration. Since the fractional volume is equal $V_{fr} = R^3/r_m^3 = d^3/8r_m^3$, then we obtain that $a_S = z^{1/2} d (12\delta V_{fr})^{-1/2}$. Substituting this value in Eq. (4), we have $d < 0.25(12\delta)^{3/2} z^{-1/2} V_{fr}^{3/2} a_{cr}^{(R)} = d_{max}$. For self-consistency, that this condition coincides with the condition $a_S < a_{cr}^{(av)}$ it is required to put $\delta = 1/3.$, that is, cracks exceeding the critical size in the bitumen $a_{cr}^{(B)}$ will grow rapidly to the size of $a_S = z^{1/2} d (12\delta V_{fr})^{-1/2}$, until stopped by modifier particles. After this the rapid crack growth takes turns to the slow fatigue growth from the size a_S to a critical size in averaged medium $a_{cr}^{(av)} = V_{fr} a_{cr}^{(R)}$. In the calculations the fatigue crack growth was used Paris' law:

$$da / dN = A(\Delta K)^n ; K = \sigma_{loc} \sqrt{\pi a} \tag{5}$$

where A and n are material parameters, ΔK is the amplitude the stress intensity factor change. Fate of crack can be as follows. The initial embryonic submicrocracks of size a_{ini} grow slowly and then exceed the size $a_{cr}^{(B)}$. Thereafter, they are rapidly super critically growing in the bitumen to a size $a_S (d)$, until stopped by modifier particles. After stopping by modifier particles cracks will grow by fatigue way to the critical size a_{cr}^{av} of averaged medium. The two parallel process of cracks conversion in the supercritical are possible.

1. Accumulation of cracks at the boundary $a_s(d)$ and their subsequent merging to generate a supercritical crack. Let the concentration of initial submicrocracks is ac_0. As a result of cyclic impact they transform into supercritical state for bitumen matrix in characteristic time intervalt. Then they propagate fast in the matrix to the size $a_s(d)$ and then stop. Concentration c of the stopped cracks varies as $dc/dt=ac_0/t$. At a concentration of $c = c_{max} - a_{stop}^{-3}$ the stopped cracks begin to touch each other and then merge to generate a large supercritical crack which can turn into a trunk crack, destroying the entire sample. The value t can be estimated as durability of the pure bitumen material.

2. The second process is fatigue growth of stopped crack from the length of $a_0 = a_s = 0.5z^{1/2} V_{fr}^{-1/2}d$ to a length $a = a_{cr}^{(av)}$. Assuming that the fatigue crack growth occurs under the Paris law (5) the ratio N_r of modified bitumen durability (measured by the number of load cycles) and of unmodified bitumen durability was calculated. The results of our calculations using of Eqs. (2)–(5) are shown in Fig. 3.2.

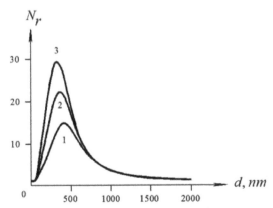

FIGURE 3.2 Calculated dependences of the relative durability N_r of cyclic loaded rubber-bitumen binder samples on the sizes of the rubber particles introduced into the bitumen. **Curve 1** – at an average mechanical stress in the pavement $\sigma = 2.5$ MPa. **Curve 2** ($\sigma = 3.75$ MPa) Curve 2 ($\sigma = 5$ MPa). The relative volume of rubber inclusions in rubber-bitumen binder $V_{fr} = 5\%$.

3.3 CONCLUSION

As can be seen from Fig. 3.2, it is necessary to use very small (150–750 nm) rubber particles in the asphalt binder to ensure optimal high durability of pavement. It should be noted that obtaining of the most inexpensive raw materials (tire rubber fine particles) for successful modifying bitumen was possible only by using the original method of "High-temperature shear-induced grinding of polymers and their composites" developed by Prof. Nikolskii. Recently, it was shown that the durability of bitumen binder samples manufactured with using rubber particle sizes (about 0.5–0.8 microns) is several times greater the durability of samples manufactured from known domestic and foreign brands of bitumen binders modified by rubber with using rubber particle sizes higher than 2–4 micron [11].

ACKNOWLEDGMENTS

The authors thank Prof. V. G. Nikol'skii for valuable discussions and Prof. G. E. Zaikov for support of this study.

KEYWORDS

- **Elastomer-bitumen binder**
- **Micro-cracks**
- **Pavement**
- **Tire crumb rubber**
- **Trunk cracks**

REFERENCES

1. Kaplan, A. M., & Chekunaev. N. I. (2010). *Theoretical Foundations of Chemical Engineering, 44(3)*, 339–347.
2. Proceedings "7th RILEM International Conference on Cracking in Pavements," edited by Scarpfs, A., Kringos, N., Al-Qadi, I., Loizos Dordrecht, A., & Heidelberg, NY. (2012). London. Springer Publisher, 1378pp.
3. Cheremskoy, P. G., Slezov, V. V., & Betekhin, V. I. (1990). *Pores in solids* Energoatomizdat, Moscow (in Russian).

4. Rudenskii, A. V. (1992). Road asphalt pavements. Moscow, Transport (in Russian).
5. Bucknall, C. B. (2000). In *"Polymer blends,"* Paul, D. R. & Bucknall, C. B., (Eds.) John Wiley & Sons Inc., Toronto, NY.
6. Georg, H., & Michler. (2008). *Electron microscopy of polymers*. Springer.
7. Elliott, H. A. (1947). An analysis of the conditions for rupture due to Griffith cracks, *Proc. Phys. Soc.*, *58*, 208–223.
8. Hesp, M. (2004). Development of a Fracture Mechanics-Based Asphalt Binder Test Method for Low Temperature Performance Prediction. Final Report for Highway IDEA Project 84. Transportation Research Board of the National Academies.
9. Al-Quraishi, A. A. (2007). *The Deformation and Fracture Energy of Natural Rubber* under high strain rates. PhD, University of Akron.
10. Chekunae, N. I. & Kaplan, A. M. (2011). *Key Engineering Materials*, 462–463, 506–511.
11. ISSN 1990_7931, *Russian Journal of Physical Chemistry B, 8(4)*. © Pleiades Publishing, Ltd. (2012). Original Russian Text © Nikol'skii, V. G. (2014). At all will be published in *Khimicheskaya Fizika, 33(7)*.

CHAPTER 4

A CASE STUDY ON THE STRUCTURE AND PHYSICAL PROPERTIES OF THERMOPLASTICS

MARIA RAJKIEWICZ, MARCIN ŚLĄCZKA, and JAKUB CZAKAJ

CONTENTS

ABSTRACT

The structure and physical properties of the thermoplastic vulcanizates produced in the process of the reactive processing of polypropylene and ethylene-octane elastomer in the form of alloy, using the cross-linking system was analyzed.

4.1 INTRODUCTION

With the DMTA, SEM and DSC it has been demonstrated that the dynamically produced vulcanizates constitute a typical dispersoid, where semi-crystal PP produces a continuous phase, and the dispersed phase consists of molecules of the cross-linked ethylene-octane elastomer, which play a role of a modifier of the properties and a stabilizer of the two-phase structure. It has been found that the mechanical as well as the thermal properties depend on the content of the elastomer in the blends, exposed to mechanical strain and temperature. The best results have been achieved for grafted/cross-linked blends with the contents of iPP/EOE-55/45%.

Three com. ethene/n-octene copolymers (I) were blended with a com. isotactic polypropylene (II) grafted/crosslinked with mixts. Of unsaturated silanes and $(PhCMe_2O)_2O$ under conditions of reactive extrusion at 170–190 °C and studied for mech. Properties, thermal stability and microstructure. The composite materials consisted of semi crystals II matrix and dispersed small pertides of cross-linked I. The best mech. properties were when the II/I mass ratio was 55/45.

The thermoplastic elastomers (TPE) are a new class of the polymeric materials, which combine the properties of the chemically cross-linked rubbers and easiness of processing and recycling of the thermoplastics [1–8]. The characteristics of the TPE are phase micrononuniformity and specific domain morphology. Their properties are intermediate and are in the range between those, which characterize the polymers, which produce the rigid and elastic phase. These properties of TPE, regardless of its type and structure, are a function of its type, structure and content of both phases, nature and value of interphase actions and manner the phases are linked in the system.

The progress in the area of TPE is connected with the research oriented to improve thermal stability of the rigid phase (higher T_g) and to increase chemical resistance as well as thermal and thermo-oxidative stability of the elastic phase [2]. A specific group among the TPE described in the literature and used in the technology are microheterogeneous mixes of rubbers and plastomers, where the plastomer constitutes a continuous phase and the molecules of rubber dispersed in it are cross-linked during a dynamic vulcanizing process. The dynamic vulcanization is conducted during the reactive mixing of rubber and thermoplast in the smelted state, in conditions of action of variable coagulating and stretching stresses and of high coagulating speed, caused by operating unit of the equipment. Manner of producing the mixtures and their properties as well as morphological traits made them be called thermoplastic vulcanizates (TPE-V) [9, 10]. They are a group of "customizable materials" with configurable properties. To their advantage is that most of them can be produced in standard equipment for processing synthetics and rubbers, using the already available generations of rubbers as well as generations of rubbers newly introduced to the market with improved properties. A requirement of developing the system morphology (a dispersion of macromolecules of the cross-linked rubber with optimum size in the continuous phase of plastomer) and achieving an appropriate thermo plasticity necessary for TPE-V are the carefully selected conditions of preparing the mixture, (temperature, coagulation speed, type of equipment, type and amount of the cross-linking substance). When selecting type and content of elastomer, the properties of the newly created material can be adjusted toward the desirable direction. Presence of the cross-linked elastomer phase allows for avoiding glutinous flow under the load, what means better elasticity and less permanent distortion when squeezing and stretching the material produced in such a manner as compared to the traditional mixtures prepared from identical input materials, each of which produces its own continuous phase. With the dynamic vulcanization process many new materials with configured properties have been achieved and introduced to the market. The most important group of TPE-V, which has commercial significance, are the products of the dynamic vulcanization of isotactic polypropylene (iPP) and ethylene-propylene-diene elastomer (EPDM). It is a result of properties of the PP and EPDM system, which, due to presence of the double bonds, may be cross-linked with conventional systems, which are of relatively low price, good contents miscibility and ability to be used within the temperature range 233–408 K [11–15].

Next level in the field of thermoplastic elastomers began with development of the metallocene catalysts and their use in stereo block polymerization of ethylene and propylene and copolymerization of these olefins with other monomers, leading to macromolecules with a "customized" structure with a microstructure and stereo regularity defined upfront. The catalysts enabled production of the homogeneous olefin copolymers, which have narrow distribution of molecular weights ($RCC=M_w/M_n<2.5$). According to the developed technology called Insite and using the on-place catalyst [16, 17]. The Dow Chemical Co-company produces the olefin elastomers Engage™, which contain over 8% of octane. Co-polymers Engage are characterized by lack of relation between the traditional Mooney viscosity and the technological properties. Compared to other homogeneous polymers with the same flow index, they are characterized by higher dynamic viscosity at zero coagulation speed and decreasing viscosity at increasing coagulation speed. They have no fixed yield point. Saturated nature of elastomer, caused by absence of diene in the chain, results in some restrictions in choice of the cross-linking system. The ethylene-octane elastomers can be easily radiation cross-linked with peroxides or moisture, if they are formerly grafted with silanes. There is relatively not much description of the behavior of the ethylene-octene elastomer in the dynamic vulcanization process in the literature. The specific physical properties of such elastomers and possibility to process them within a periodic process as well as within a continuous process, due to convenient form of the commercial product (granulate) encouraged us to start the recognition works on development of the technology of producing thermoplastic vulcanizates from the mixture of elastomer Engage and iPP.

Use a silane-based cross-linking system in the dynamic vulcanization process seemed the most interesting. For the research works one of the known methods of cross-linking polyolefins with silanes was used, assuming that the cross-linking of EOE would proceed according to the analogous mechanism. In the seventies of the twentieth century the Dow Corning Co. company developed two methods of the hydrolytic cross-linking of polyolefins grafted with vinylosilanes according to the radical mechanism [18, 19]. Nowadays, three polyolefin cross-linking methods are widely used in the industrial production. The grounds for distinguishing them are technological equipment and procedure. It is one-phase and two-phase method and a "dry silane method," available only under license

[20]. The mechanism of cross-linking PE with the cross-linking system: silane/peroxide/moisture is shown schematically in Fig. 4.1.

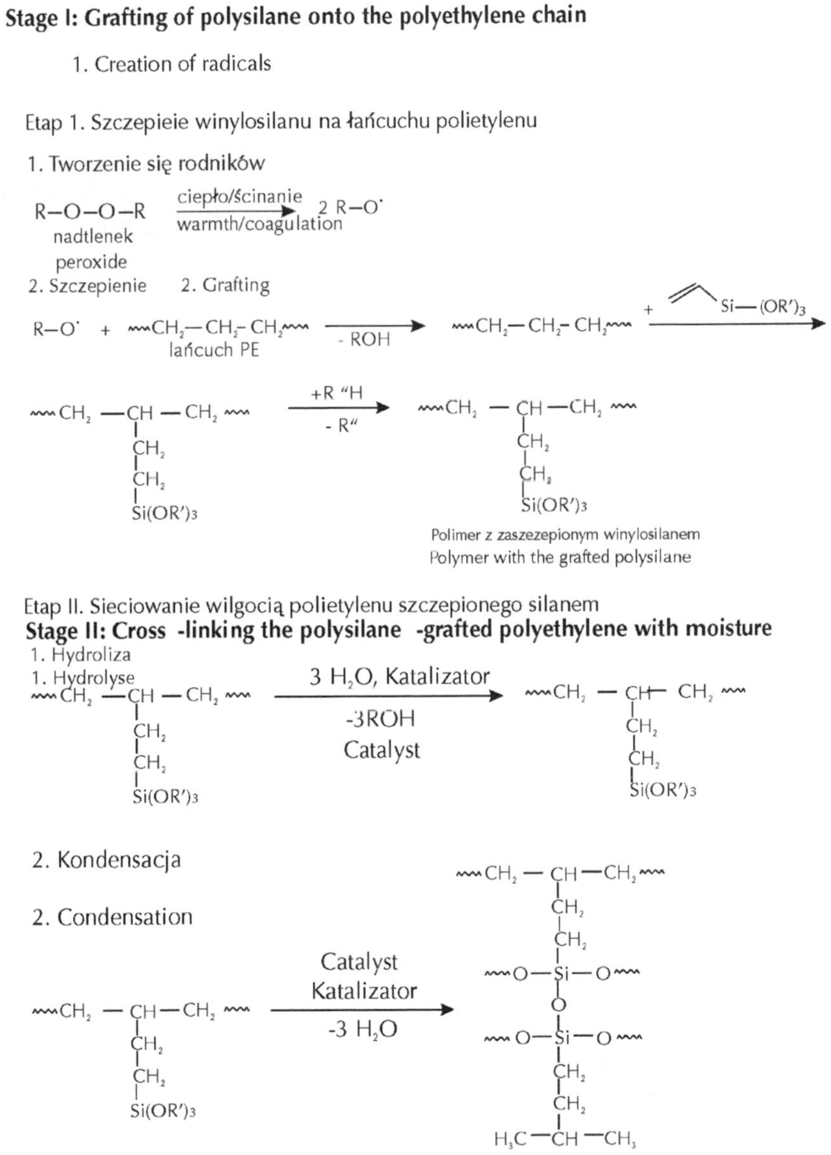

Stage I: Grafting of polysilane onto the polyethylene chain

1. Creation of radicals

Etap 1. Szczepieie winylosilanu na łańcuchu polietylenu

1. Tworzenie się rodników

$$R-O-O-R \xrightarrow[\text{warmth/coagulation}]{\text{ciepło/ścinanie}} 2\ R-O^{\cdot}$$

nadtlenek
peroxide

2. Szczepienie   2. Grafting

$$R-O^{\cdot} + \text{\textasciitilde\textasciitilde}CH_2-CH_2-CH_2\text{\textasciitilde\textasciitilde} \xrightarrow{-ROH} \text{\textasciitilde\textasciitilde}CH_2-CH_2-CH_2\text{\textasciitilde\textasciitilde} + \diagup\hspace{-0.3em}\diagup Si-(OR')_3 \longrightarrow$$

łańcuch PE

$$\text{\textasciitilde\textasciitilde}CH_2-CH-CH_2\text{\textasciitilde\textasciitilde} \xrightarrow[-R'']{+R''H} \text{\textasciitilde\textasciitilde}CH_2-CH-CH_2\text{\textasciitilde\textasciitilde}$$

with side chains:
$$\begin{array}{c} | \\ CH_2 \\ | \\ CH_2 \\ | \\ Si(OR')_3 \end{array}$$

Polimer z zaszezepionym winylosilanem
Polymer with the grafted polysilane

Etap II. Sieciowanie wilgocią polietylenu szczepionego silanem
Stage II: Cross -linking the polysilane -grafted polyethylene with moisture
1. Hydroliza
1. Hydrolyse

$$\text{\textasciitilde\textasciitilde}CH_2-CH-CH_2\text{\textasciitilde\textasciitilde} \xrightarrow[\substack{-3ROH \\ \text{Catalyst}}]{3\ H_2O,\ Katalizator} \text{\textasciitilde\textasciitilde}CH_2-CH-CH_2\text{\textasciitilde\textasciitilde}$$

side chains $CH_2-CH_2-Si(OR')_3$

2. Kondensacja

2. Condensation

$$\text{\textasciitilde\textasciitilde}CH_2-CH-CH_2\text{\textasciitilde\textasciitilde} \xrightarrow[-3\ H_2O]{\substack{Catalyst \\ Katalizator}}$$

side chains $CH_2-CH_2-Si(OR')_3$

Product structure:
$$\text{\textasciitilde\textasciitilde}CH_2-CH-CH_2\text{\textasciitilde\textasciitilde}$$
$$| \quad CH_2$$
$$CH_2$$
$$\text{\textasciitilde\textasciitilde}O-Si-O\text{\textasciitilde\textasciitilde}$$
$$O$$
$$\text{\textasciitilde\textasciitilde}O-Si-O\text{\textasciitilde\textasciitilde}$$
$$CH_2$$
$$CH_2$$
$$H_3C-CH-CH_3$$

FIGURE 4.1 Crosslinking mechanism of polyolefins with a silane. (a) *Rigid bond C-C*; (b) *Elastic bond Si-O-Si.*

The process of catalytic hydrolyzes of the alcoxylene groups of the grafted silane to the silane groups and, then, the catalytic condensation of the silane groups leads to production of the cross-linked structure through the siloxane groups. The hydrolyze and the condensation take place in an increased temperature with presence of the catalyst and water. Dibutyl tin dilaurate (DBTL) is most often used as a catalyst of the reaction. The catalyst may be added either to the polymeric blend (it constitutes an increased risk of the premature cross-linking), or in the form of a premixed reagent during the processing. The mechanism of action of DBTL, as a cross-linking catalyst, is complex and has not been sufficiently explained.

As a result of cross-linking the polyolefins with silanes the Si-O-Si bonds are produced, which are more elastic than the rigid bonds C-C created as a result of cross-linking of polymers induced by radiation and peroxides (Fig. 4.2). Use of silanes gives more elastic products and the cross-linking process is more cost-effective.

Sztywne wiązanie c-c

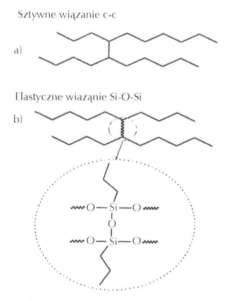

FIGURE 4.2 Structure of polyolefins crosslinked with a (a) peroxide or radiation, (b) with a silane.

4.2 EXPERIMENTAL PART

4.2.1 *RAW MATERIALS*

Isotactic polypropylene Malen P-F401 iPP, for extrusion, made by Orlen SA; flow index 2.4–33.2 g/10 min, yield point in stretching 28.4 MPa, crystallinity level 95%;

- The ethylene-n-octene elastomer s EOE type Engage, synthesized according to the Insite technological process, manufactured by Du-Pont Dow Chemical Elastomers (Table 4.1);
- Silanes: Silquest A-172 vinylo-tris (2-methoxyethoxysilane), Silquest A-174–3-methacryloxypropyltrimethoxysilane, manufactured by Vitco SA;
- Dicumyl peroxide with 99% content of the neat peroxide, manufactured by ELF Atochem;
- Antioxidant tetra-kis (3, 5-di-tetra-butyl-4-hydroxyphenyl) propionate, manufactured Ciba-Geigy.

4.2.2 TEST METHOD

Three types of EOE from the wide range offered by the manufacturer were selected (Table 4.1). The general-purpose elastomers were selected with high content of octane and a defined characteristic.

The test were made aimed for determining a threshold value of content of elastomer in the iPP/EOE mixture, which was subject of the dynamic vulcanization process, considering the influence of these parameters on variable properties of iPP. A series of tests was made, in which the proportions of PP and elastomer Engage I were changed in the range 15–60%, with continuous addition of the cross-linking system (silane A–172/dicumyl peroxide) 3/0.03% in relation to elastomer and antioxidant additive 0.2%.

For the tests of preparing dynamic vulcanizates in the continuous process of reactive extrusion a twin-screw mixer-extruder DSK 42/6D manufactured by Brabender was used. The vulcanizates were produced dynamically in the process of one-stage or two-stage extrusion process, setting the favorable operating parameters for the device, which had been determined based on multiple tests: distribution of temperatures in each heating area of the extruder: 170/180/190 °C, screw rotation: 40/min. In the one-stage process all the components provided in the formula (elastomer, iPP, anti-oxidant, silane initiating system/peroxide) was initially mixed in a fast-rotating mixer type Stephan in temperature of 50 °C, next a granulate was extruded. In the two-stage process in the first stage the iPP, elastomer and antioxidant mixture was extruded, next after mixing granulate with the cross-linking system it was extruded again.

The profiles were formed from granulates with an injection molding machine type ARBURG-420 M1000-25 All-rounder. For the tests the actual injection at speed of 10 cm^3/s was used with addition at speed of 15 cm^3/s, injection temperatures: 195/200/210/210 °C and blend injection time was slightly lower than for iPP itself.

4.2.2.1 METHODS OF ANALYZING THE BLEND

Hardness was marked according to the Shore method, scale D according to PN-ISO 868 or according to the ball insertion method according to the PN-ISO 868 (MPa). Flow speed index (MFR) was determined according to PN-ISO 1033. Resistance properties of the blend with static stretching were tested according to ISO-527, using a digital tester Instron 4505 (tear off speed: 50 mm/min). The bending properties were determined according to PN-EN ISO 178. In addition to the regular tests, the selected blends were subject to specialist examination, such as the thermogravimetric analysis (TGA), scanning electron microscopy (SEM), differential scanning calorimetry (DSC) and dynamic thermal analysis of mechanical properties (DMTA).

The samples were heated in the ambient temperature in temperature range of 30–490 °C with speed of 5 deg/min. The test was conducted with thermobalance TGA manufactured by "Perkin Elmer." Turning points were made after freezing the samples in the liquid nitrogen for about 3 min. The surfaces of the turning points were concocted with gold with vacuum powdering. The SEM JSM 6100 manufactured by JEOL was used to conduct the tests. The photographs have been made in magnification of 2000*.

TABLE 4.1 Properties of Ethylene-Octane Elastomers Engage

Elastomer type Properties	Engage I	Engage II	Engage III
Co-monomer content, % of weight (^{13}C NMR/FTR)	42	40	38
Density, g/cm^3, ASTM	0.863	0.868	0.870
Mooney viscosity, ML (1+4) 121 °C	35	35	8
MFR, deg/min, ASTM D-1238	0.5	0.5	5.0
Shore hardness A, ASTM D-2240	66	75	75

TABLE 4.2 Selected Properties of the Dynamically Cross Linked Blends in Relations to PP/EOE Ratio

Ratio PP/Engage I, % of weight	100/0	85/15	70/30	55/45	40/60
Hardness, °ShD	80	63	57	50	36
MFR (190 °C, 2, 16 kg), g/10 min	2.4	1.63	1.29	1.28	1.15
MFR (190 °C, 5), g/10 min	–	5.06	5.89	5.80	4.90
$T_{A\,120}$, °C*	152	143	130	106	~60
Hardness HK, MPa**	24.7	16.1	12.2	11.8	8.7
Solubility of elastomer in cyclohexane, %	—	—	13.9	12.03	14.2
Solubility of elastomer <t4/> in boiling xylene, %	—	—	24.0	33.0	42.0

*T_{A120} – Vicat softening point,
**HK – ball pan hardness method.

4.2.3 RESULTS OF THE TESTING

Influence of the content of elastomer Engage I on physical properties of TPE with PP and EOE modified (grafting/cross-linking) with a silane/peroxide cross-linking system has been shown in Fig. 4.3 and Table 4.2. The content of comonomer had significant influence on such properties of the elastomer as elasticity, modulus, density and hardness. Values of two last parameters decreased with the increase of content of *n*-octene in elastomer. It has been stated that properties of the dynamically vulcanized blends could be adjusted with content of the elastomer phase. With the increase of content of EOE in range 15–45% tensile strength increased (18–30 MPa), and, in the same time, relative elongation increased with tear off (300–700%). With elastomer content over 50% a visible decrease of both properties occurred, which came to 15 MPa and 600%, respectively. Whereas hardness expressed in Sh degrees or in MPa) systematically decreased with the increase of the content of EOE in the blend. The optimum content of EOE introduced to PP was 45% and therefore in most subsequent tests a blend was used, in which iPP/EOE ratio was 55/45%. Such contents had also the blends listed in Table 4.3, made of three types of EOE and two types of silane, with constant content of the cross-linking system (silane A–174/dicumyl peroxide 3.0/0.01%, Irganox 1010–0.2%. The blends containing elastomers Engage I and Engage II, with difference of content of octane by 2% Shore hardness A (66 and 75, respectively) and with very similar Mooney viscosities, showed comparable resistance and rheological characteristic. The blends containing elastomer Engage III, with the lowest octane content, were characterized by slightly lower variables of tensile strength (tension at the tear off), elongation and hardness, but by a much higher tension at the yield point and high flow index.

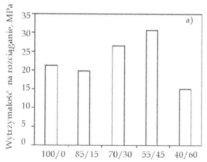 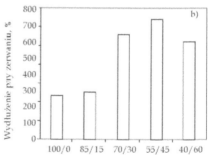

FIGURE 4.3 Mechanical properties of the dynamically cross-linked blends in relation to PP/EOE ratio, (a) tensile strength, (b) elongation at the tear off.

Such behavior of elastomer Engage III blended with PP resulted probably from its different rheological characteristic, including its four times lower viscosity and very high flow index as compared to Engage I, which was recognized as the most suitable for production of nonsaturated blends using the dynamic vulcanization method.

TABLE 4.3 Effect of Type of Elastomer Engage Modified with Silane A-174 on the Properties of the Dynamically Cross-Linked PP/EOE–55/45% Blends

Elastomer	Engage I	Engage II	Engage III
Blend properties			
MFR, g/10 min	1.86	1.80	4.07
(2,16 kg, 190 °C)			
Gardbess, °Sh, D	42/39	42/40	39/38
$\epsilon_{B,}$ %	720	752	660
$\sigma_{M,}$ MPa	25.2	29.5	21.9
$\sigma_{100\%,}$ MPa	11.1	12.6	11.0
σ_{y}, MPa	11.1	12.6	11.1
ϵ_{y}, %	24.0	27.9	39.9

Symbols; $\epsilon_{100\%}$ – tension at 100% elongation, σ_{M} – maximum tension, ϵ_{B} –relative elongation at the tear off, σ_{y} –yield point, ϵ_{y} –elongation on yield point.

Blend with the selected optimum contents iPP/Engage I 55/45% and the selected cross-linking system (silane/peroxide 3/0.03%) were characterized by high thermal stability, independent from type of the material employed to cross-linking silane. It has been confirmed with tests of TGA of blend containing silane A-172 and silane A-174 (samples PL-1 and PL-2, respectively), what is shown in Table 4.4 in temperature of 230 °C the decrease of weight did not exceed 0.5%.

TABLE 4.4 Results of the Thermogravimetric Analysis of iPP/EOE-55/45% Blend (Engage I)

PL-1(Silane A–172)		PL-2 (Silane A–174)	
Temperature, °C	Decrease of weight, %	Temperature, °C	Decrease of weight, %
230	0.23	230	0.36
300	7.43	300	7.66
352	25.97	378	54.36
363	40.06	405	89.63
430	94.05	426	94.36

In 300 °C temperature came to as much as 7.5%, and the further increase of temperature caused the progressive degradation process. Analysis of the morphological structure of grafted/cross-linked iPP/Engage I blend using the SEM, DSC and DMTA methods showed that the blends produced with dynamic vulcanization had a special two-phase structure. With scanning electron microscopy photographs of surface of turning points of iPP samples and iPP/Engage I 55/45% blends have been made (Fig. 4.4). The SEM analysis showed that the obtained blends were mixtures of two thermodynamically nonmiscible structures. The continuous phase of iPP had a visible semicrystal structure and the spherical and oval molecules of the dispersed phase of the cross-linked elastomer were not connected to the continuous phase. The viscoelastic properties, assessed with the DMTA Mk II equipment manufactured by Polymer Laboratories in the sinusoidally variable load conditions at bending with frequency of 1 Hz, in the temperature range between −100 and +100 °C also showed heterogeneous structure of the produced blends.

 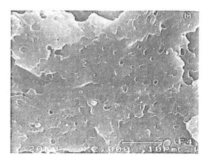

FIGURE 4.4   SEM microphotographs of: (a) neat iPP, (b) dynamically cross-linked PP/EOE blend – 55/45%; magnification 2000×.

In Fig. 4.5, course of changes of the storage modulus E,' loss modulus E" and vibration damping factor gδ for iPP and iPP/Engage I blends with content of 85/15, 70/30 and 55/45% in relation to temperature has been shown. For iPP/EOE blends two, clear relaxation transitions in the range of glass transition are visible, near glass points of iPP and EOE.

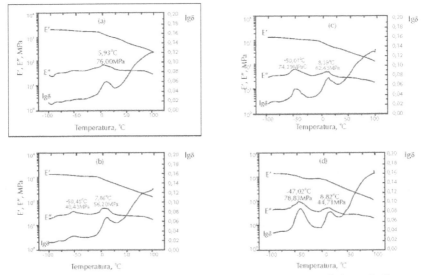

FIGURE 4.5 The dynamic mechanical properties of neat PP and dynamically cross-linked PP/EOE blends in relation to temperature: storage modulus (E'), loss modulus (E"), loss tangent (tanδ); (a) PP (b) PP/EOE 85/15, (c)-PP/EOE 70/30, (d)PP/EOE 55/45.

Addition of elastomer slightly moved the glass point of iPP toward higher temperatures. PP showed higher values of the E' modulus as compared to the analyzed composites, whereas in the chart E" one maximum appeared corresponding to T_g PP.

TABLE 4.5 Selected Properties of PP and Dynamically Cross-Linked PP/EOE Blend PP/EOE (55/45%) in the Reactive Processing

Properties	PP	TE-1 (Silane A-174)	TE-2 (Silane A-172
Yield point, MPa	31.4	12.1	12.5
Relative elongation of yield point, %	9.4	25.1	33.0
Tensile strength, MPa	15.1	15.5	18.3
Elongation at the tear off, %	196	443	440
Tensile shear modulus, MPa	1455	476	430
Bending strength, MPa	39.6	13.0	10.4
Tensile bending modulus, MPa	1499	504	415
Young modulus, MPa	1520	515	502
HDT, load 1.8 MPa, °C	50.5	38	38
Izod notched impact strength, kJ/m^2	3.16	46.7	43.0

On the DSC thermal images made in positive temperatures (50–210 °C) a visible maximum appeared which was connected with thermal transition corresponding to the melting point of iPP. Systematic decrease of melting point of the thermoplastic phase of iPP in iPP/Engage I blends related to the original polymer was observed (Fig. 4.6). Causes of these changes could not be unambiguously determined – it is supposed that here the phenomena such, as degradation of iPP in conditions of the high-temperature processing change of semicrystal structure of iPP may have occurred. In order to compare the properties of iPP/Engage I blend with content of 55/45%, produced with periodical method and in the one-stage or two-stage continuous process, a series of tests with use of the general formula was performed. Properties of the selected blends cross-linked with the silane A-174/dicumyl peroxide (TE-1) and silane A-172/dicumyl peroxide (TE-2) systems have been listed in Table 4.5.

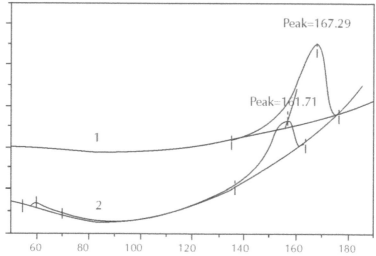

FIGURE 4.6   DSC curves of (1) PP; (2) PP/EOE dynamically cross-linked 55/45%.

4.3 CONCLUSIONS

The conducted tests leaded to developing grounds for the technology for dynamic vulcanization of materials with thermo-elastoplastic properties, in which a thermoplastic polymer constitutes a continuous phase, whereas the dispersed phase consists of cross-linked elastomer particles. Basic elastomers are polyisoprene with isotacticity level of 85% or higher and copolymer EOE containing over 30% of n-octane.

The blend properties can be adjusted with content of the elastomeric phase and with cross-linking silane/peroxide system. The achieved material has a heterogeneous structure and favorable set of resistance properties, including higher Izod notched impact strength as compared to PP.

The developed technology allows for achieving new materials with preconfigured properties, competitive to the unmodified iPP and to physical PP/elastomer mixtures. They could be processed in the equipment for synthetics. These blends may be used as structural materials, characterized by higher thermal resistance as compared to the unmodified PP. They may also be a generation of modifiers for polyolefins and polymeric mixtures.

KEYWORDS

- Copolymer
- Physical PP/elastomer mixtures
- Polyolefins and polymeric mixtures
- Synthetics
- Thermal resistance
- Thermo-elastoplastic properties

REFERENCES

1. Rzymski, W., & Radusch, H. J. (2002). *Polimery, 47*, 229.
2. Spontak Richard, J., & Patel Nikunj, P. (2000). *Current Opinion in Colloids and Interface Sci., 5,* 333.
3. Rzymski, W., & Radusch, H. J. (2005). *Polimery, 50*, 247.
4. Radusch, H. J., Dosher, P., & Lohse, G. (2005). *Polimery. 50*, 279.
5. Rzymski, W. M. (1998). *Stosowanie i przetwórstwo materiałów polimerowych* Wyd. Polit. Częstochowskiej, Częstochowa, 17–28.
6. Holden, G. (1991). *Understanding Thermoplastic Elastomers,* Hanser Publishers Munich, (2000).
7. Rader, C. P. *Modern Plastic Encyclopedia.*
8. Rader, C. P. (1993). *Kuststoffe, 83*, 777.
9. Rzymski, W., & Radusch, H. J. (2001). *Elastomery, 5(2)*, 19.
10. Rzymski, W., & Radusch, H. J. (2001). *Elastomery, 5(3)*, 3.

11. Winters, R. R. (2001). *Polimery* 42, 9745.
12. Trinh an Huy, T., Luepke, H. J., & Radusch. (2001). *App., Polym., Sci.*, *80,* 148.
13. Jain, A. K., Nagpal, A. K., Singhal, R., & Gupta Neeraj, K. (2000). *J. Appl. Polym. Sci.*, *78*, 2089.
14. Gupta Neeraj, K., Janil Anil, K., Singhal, R., & Nagpal, A. K. (2000). *J. Appl. Polym. Sci., 78*, 2104.
15. Suresh Chandra Kumar, S., Alagar, M., & Anand Prabu, A. (2003). *Eur. Polym. Journal, 39*, 805.
16. Fanicher, L., & Clayfield, T. (1997). *Elastomery, 1,* 4.
17. ENGAGE polyolefin elastomers. (2003). A Product of Du Pont Dow Elastomers Product Information.
18. Voight, H. U. (1981). *Kautsch. Gum. Kunstst., 34*, 197.
19. Toynbee, J. (1994), *Polymer 35*, 428.
20. Special Chem. Crosslinking Agent Center, Dane techniczne (2004).

CHAPTER 5

AMYLOSE DESTRUCTION AND FREE RADICALS GENERATION UNDER SHEAR DEFORMATION

S. D. RAZUMOVSKII, V. A. ZHORIN, V. V. KASPAROV, and A. L. KOVARSKI

CONTENTS

ABSTRACT

Amylosa degradation under shear deformation at high pressure has been studied using Bridgman anvils. The range of pressure was 25 MPa and the anvils turn angle 520 degrees. The processed samples have been studied by electron spin resonance spectroscopy, viscosimetry, weight analysis, ozone treatment. Processing has been accompanied by scission of amylose macromolecules, double bonds formation and water molecules detachment. Molecular weight decreased about two times. It has been established that ESR spectra consist of two lines one of which corresponds to amylose free radicals and another to ultra dispersed ferromagnetic particles incorporated into the sample from the anvils. The number of stabilized free radicals was 0.3–0.4% of the total number of chain scissions. The basic ways of the free radicals stabilization probably are the delocalization of the spin density of unpaired electron over the system of conjugated double bonds and low molecular mobility of amylose.

5.1 INTRODUCTION

In most processes of conversion of the plant biomass into useful products the first stage is its mechanical grinding [1, 2]. The grinding of biomass may cause various chemical reactions induced by force fields. The nature of these processes and how to control them are not well understood up to the present. Previously it was shown that mechanical deformation may accelerate hydrolytic processes in biomass [3], contribute to delignification of lignocelluloses materials [4] and can lead to the formation of free radicals (FR) [5–7]. Strong strain leads to demolition of the carbohydrate frame of the material and complex sets of products are formed [8]. Ways of their formation can be understood by exploring low-molecular carbohydrate fragments and the nature of polysaccharides macromolecules end groups. It is clear that the destruction of polysaccharides macromolecules under mechanical load must be accompanied, first of all, by the glycosidic bonds rapture and FR formation [5–7, 9].

In this paper we have studied amylose as a model carbohydrate, which is one of the typical representatives of polysaccharides, constituent part of starch [10]. It is convenient to use because it has a small molecular weight and its ability to dissolve in widespread solvents, including hot water.

The objective of the study was to obtain some quantitative data on the amount of FR formed in the process of shear deformation of amylose under high pressure. Thereto we used electron spin resonance spectroscopy (ESR), method allowing identifying FR even in small quantities [11]. In addition, we considered it necessary to get the data about the changes of amylose molecular mass and on products of destruction, appearing during mechanical processing. For these purposes we used viscosimetry and weight analysis. This resulted in obtaining information on the amount of FR, macromolecules degradation degree of amylose, and the appearance of a new product the water, along with previously diagnosed oligosaccharides [12].

5.2 EXPERIMENTAL PART

We have studied highly purified amylose derived from potato tubers (SERVA, Germany). Average molecular mass of amylose was 35°000. Mechanical treatment of samples has been made using hardened steel Bridgman anvils [8, 13] with a diameter of 20 mm. The anvils were placed in a hydraulic press. Treatment of the sample (30 mg) has been carried out at pressures of 25 MPa. After 2 min exposure the anvils have been turned relative to each other at an angle of 520degrees.

Spectra of electron spin resonance (ESR) were recorded using ESR-spectrometer "Bruker EMX" in the X-wavelength range. Viscosity of amylose samples before and after treatment was measured by capillary viscosimeter previously dissolved them in dimethyl sulfoxide (0.6 g of polymer or products in 10 cm^3 of DMSO).

The amount of water formed was determined by the weight loss after drying of the sample in vacuum at 60°C during 2 h at 10 mm of Hg.

The number of unsaturated fragments in macromolecules of destructed samples was determined according to the volume of absorbed ozone by the method described in [14].

5.3 RESULTS AND DISCUSSION

In the course of the processing the sample underwent considerable changes. The powder turned into a monolithic pill which crushing for further research required some effort. Viscosimetric measurements showed that the action of the pressure combined with shear deformation led to a halving of average molecular mass of amylose from 35,000 to 15,000 and a significant number of low-molecular oligosaccharides arised [12]. There was also the formation of water in the amount of 0.3–0.4% by weight of the sample.

All samples after mechanical treating gave ESR signals (Fig. 5.1). As seen from Fig. 5.1, the ESR spectrum consists of two lines: broad and narrow. The width of broad line is $\Delta H_1=730G$, and narrow one is $\Delta H_2=12.3G$. The values of g-factor are $g_1=2.2219$ and $g_2=2.0034$. Both signals are rather stable and when rerecording of the spectrum after exposure within a month at room temperature no changes in the parameters of the spectra were observed. After the sample dissolution in water and subsequent drying narrow singlet disappeared but wide remained.

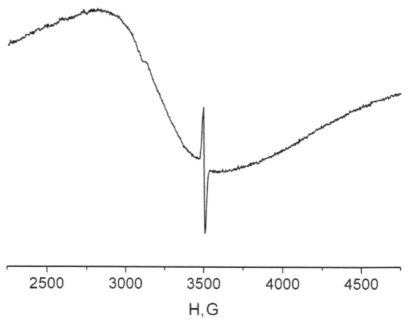

2500   3000   3500   4000   4500

H, G

FIGURE 5.1   ESR spectrum of amylose subjected to shear deformation under high pressure.

To understand better the nature of the observed ESR signals additional experiments were performed. Amylose powder ground in an agate mortar, and then ESR spectrum was analyzed. In this series of experiments only narrow line was observed. Broad line was absent.

Preservation of broad lines after the sample dissolution in water and their disappearance when grinding the sample in an agate mortar suggests that broad line appears due to ferromagnetic ultra dispersed particles scraped off the steel anvils surface. It was shown earlier that broad ESR lines of several hundred Gauss are typical for ferromagnetic nanoparticles and not characteristic of organic radicals [15–17]. Narrow lines observed in the ESR spectra can be assigned to unpaired electrons formed at amylose macromolecules rapture in the process of shear deformation. Concentration of free radicals was determined by double integration of ESR spectra of amylose radical and standard sample of DPPG (diphenylpicrylhydrazyl). The concentration of free radicals in the sample was $\sim 10^{-17}$ spin/g. An error in the determination of the concentration was about 50%.

Taking into consideration the value of amylose molecular mass (35,000), 1 g of the starting sample contained $(1–6.02–10^{23}/35,000) = 1.7–10^{19}$ of polysaccharide chains. After shear deformation the molecular mass decreased twice, and therefore, the number of disrupted bonds should be the same $(1.7–10^{19}$ per 1 g). Theoretically, each disrupted bond must form two free radicals that would correspond to $(3–4)10^{19}$ spins. In the experiments the number of spins found was $\sim 10^{17}$ spin/g that was 300–400 times less than the expected number.

Because the number of ruptured chemical bonds is closely related to molecular mass change the spins deficit found cannot be explained by the recombination of primary radicals. Otherwise the reduction of molecular mass would not occur. It is more likely that after the break of macromolecules their ends were particularly excited. This excitation led to the detachment of low molecular weight radicals, which interacted, with each other and with neighboring free valences. In this case the average molecular mass of the destructed sample wouldn't change essentially. However the number of free radicals decreased sharply as a result of the recombination processes in which low-molecular particles play a role of free valence carriers. A degradation of polymer materials is often accompanied by formation of low molecular weight products [18]. Water forming which we observed at the destruction of amylose was one of such products.

For analysis of the number of unsaturated fragments in macromolecules some of the samples before and after shear deformation were subjected to treatment with ozone-oxygen mixture. The amount of absorbed ozone was determined. Starting amylose practically did not absorb ozone while mechanically treated samples absorb a significant number of ozone ($8.3-10^{-8}$ mol/g). In the conditions of our experiments absorption of one molecule of ozone corresponded to the content of one C=C bond in the sample. From this it follows that absorption of ozone in amount of 8.3×10^{-8} mol that consistent with the content of $5.3-10^{16}$ double bonds per 1 g of the sample. Double bonds formation under shear strain can be explained by hydrogen atoms abstraction by end chains radicals with the subsequent relay transmission of free valences. This process, as it follows from the experiment, accompanied by the formation of water molecules.

5.4 CONCLUSIONS

Our study showed that shear deformation under high pressure of amylose, caused the formation of a relatively stable free radicals. For the first time it was shown that ESR spectrum of amylose treated by shear deformation at high pressure consists of two types of signals. One of them associated with organic free radicals, and the other with ultra dispersed ferromagnetic particles incorporated into the sample from steel Bridgman anvils. Comparison of the data on the number of radicals and decreasing of amylose molecular mass showed that the yield of the radicals was 300–400 times less than the number of ruptured macromolecules bonds. The rest of end groups of broken macromolecules stabilized apparently due to the detachment of low molecular mass fragments and water accompanied by the free valence decay [9]. By ozone treatment the number of double bonds in the samples was defined ($5.3-10^{16}$ per 1 g). It is quite likely that the delocalization of the spin density of unpaired electron over the system of conjugated double bond leads to stabilization of free radicals. It is known that such delocalization leads to blurring the hyperfine structure of ESR spectra. This is the main reason why we observe a single line of free radicals in the spectrum. Another reason is the low molecular mobility of amylose.

KEYWORDS

- Amylose
- ESR spectroscopy
- Free radicals
- High pressure
- Polysaccharides
- Shear deformation

REFERENCES

1. Mousdale, D. M. (2008). Biofuel Biotechnology, Chemistry and Sustainable Development, CRC Press, Taylor & Francis Group, Boca Rota, London, New York, 404p.
2. Moiseev, I. I., & Plate, N. A. (2006). Varfolomeev, S. D. *Herald Russ. Acad. Sci., 76,* 252–259.
3. Razumovskii, S. D., Podmaster'ev, V. V., & Zelenetsky, A. N. (2011). *Catalysis in Industry, 3,* 823–827.
4. Balashova, E. A., Salahnenko, L. S., Rogovina, S. Z., & Enikolopov, N. S. (1988). *Doklady of Academy of Science of USSR, 302,* 1134–1136 (in Rus.).
5. Abgyan, G. V., & Butyagin, P. Yu. (1965). *Vysokomol. Soed, 7,* 1410–1414 (in Rus.).
6. Shoma, J., & Sacaguchi, M. (1976). *Adv. Polym. Sci., 20,* 109–115.
7. Matveeva, N. A., Autlov, S. A., Phylinov, A. V., Bazarnova, N. G., & Lunin, V. V. (2009). *Russ. J. Phys. Chem, 83,* 860–867.
8. Kireeva, G. H., Zhorin, V. A., Razumovskii, S. D., & Varfolomeev, S. D. (2011). *Russ. J. Phys. Chem., 83,* 1187–1189 (in Engl.).
9. von Sonntag, C. (1980). *Adv. Carbohydr. Chem. Biochem., 37,* 7–77.
10. *Starch Properties and Potential* Ed. Gillard, T. (1987). Wiley and Sons: Chichester, 150p.
11. Rhodes, C. J. (2006). *Electron Spin Resonance (Some Chemical Applications),* Ann. Rep. Progr Chem., Sec. C, *102,* 166–179.
12. Kireeva, G. H., Razumovskii, S. D., & Varfolomeev, S. D. (2011). *Chemicke Listy, 105,* 730.
13. Zharov, A. A. (1994). Reaction of Solid Monomers and Polymers under Shear Deformation and High Pressure. In *High Pressure Chemistry and Physics of Polymers,* Kovarski, A. L., Ed. CRC-Press, Inc., Boca Raton, USA, 265–300.
14. Razumovskii, S. D., & Lisitsin, D. M. (2008). *Polymer Sci., Ser. A., 50,* 1187–1197.
15. Kovarski, A. L., & Sorokina, O. N. (2013). Study of Dispersion of Ferromagnetic Particles In *Update of Paramagnetic Sensors for Polymers and Composites Research,* Smiters Rapra Technology Ltd., Shawbery, UK, 91–122.
16. Kovarski, A. L., Bychkova, A. V., Sorokina, O. N., & Kasparov, V. V. (2008). *Magnetic Resonance in Solids, 10,* 25–30.

17. Ranby, B., & Rabec, J. F. (1977). ESR Spectroscopy in Polymer Research Springer-Verlag, Berlin, Germany, 332p.
18. Tatarenko, L. A., & Pudov, V. S. (1967). *Russian J. Phys. Chem., 41,* 2951–2954 (in Rus.).

CHAPTER 6

NANOELEMENT MANUFACTURING: QUANTUM MECHANICS AND THERMODYNAMIC PRINCIPLES

AREZO AFZALI and SHIMA MAGHSOODLOU

CONTENTS

ABSTRACT

Nanoscience and nanotechnology are critically important for whole of the science investigations. In this chapter, the important subjects of computational nanotechnology, namely mechanic quantum, thermodynamics and statistical mechanics and their applications in molecular systems to predict the properties and performances involving nanoscale structures and manufacturing are investigated. The main challenge for getting to the nanoelements with the most appropriate properties is the selection of manufacturing nanoelements and nanomaterials.

6.1 UNDERSTANDING OF NANOSCIENCE PRINCIPLES

A revolution is occurring in science and technology, based on the recently developed ability to measure, manipulate and organize matter on the nanoscale 1 to 100 billionths of a meter. At this level everything is attenuated to fundamental interactions between atoms and molecules. Therefore, physics, chemistry, biology, materials science, and engineering converge toward the same principles and tools. A nano element compares to a basketball, like a basketball to the size of the earth. The aim of nonscientists is to manipulate and control the infinitesimal particles to create novel structures with unique properties. The science of atoms and simple molecules and the science of matter from microstructures to larger scales are generally established, in parallel. The remaining size related challenge is at the nanoscale where the fundamental properties of materials are determined and can be engineered. A revolution has been occurring in science and technology, based on the ability to measure, manipulate and organize matter on this scale. These properties are incorporated into useful and functional devices. Therefore, nano science will be transformed into nanotechnology. Through a basic understanding of ways to control and manipulate matter at the nanometer scale and through the incorporation of nanostructures and nanoprocesses into technological innovations, nanotechnology will provide the capacity to create affordable products with dramatically improved performance. Nanotechnology involves the ability to manipulate, measure, and model physical, chemical, and biological systems at nanometer dimensions, in order to exploit nanoscale phenomena [1].

Novel properties in biological, chemical, and physical systems can be approximately obtained at dimensions between 1nanometer to 100nanometers. These properties can differ in fundamental ways from the properties of individual atoms and molecules and those of bulk materials [1]. Nowadays, advances in nanoscience and nanotechnology indicate to have major implications for health, wealth, and peace. Knowledge in this field due to fundamental scientific advances, will lead to dramatic changes in the ways that materials, devices, and systems are understood and created. Nanoscience will redirect the scientific approach toward more generic and interdisciplinary research [1, 2].

Nanoelement categories consist of atom clusters/assemblies or structures possessing at least one dimension between 1 and 100 nm, containing 10^3–10^9 atoms with masses of 10^4–10^{10} Daltons. Nanoelements are homogenous, uniform nanoparticles exhibiting well-defined (a) sizes, (b) shapes, (c) surface chemistries, and (d) flexibilities (i.e., polarizability). Typical nanoelement categories exhibit certain nanoscale atom mimicry features such as (a) core-shell architectures, (b) predominately (0–D) zero dimensionality (i.e., 1–D in some cases), (c) react and behave as discrete, quantized modules in their manifestation of nanoscale physicochemical properties, and (d) display discrete valencies, stoichiometry's, and mass combining ratios as a consequence of active atoms or reactive/passive functional groups presented in the outer valence shells of their core-shell architectures. Nanoelements must be accessible by synthesis or fractionation/separation methodologies with typical monodispersities [90% (i.e., uniformity) [3] as a function of mass, size, shape, and valency. Wilcoxon et al. have shown that hard nanoparticle Au nanoclusters are as monodisperse as 99.9% pure (C) [3]. Soft nanoparticle dendrimers are routinely produced as high as generation =6–8 with polydispersities ranging from 1.011 to 1.201 [4–6]. Nanoelement categories must be robust enough to allow reproducible analytical measurements to confirm size, mass, shape, surface chemistries, and flexibility/ polarizability parameters under reasonable experimental conditions.

TABLE 6.1   The Importance of Scales

Length (m)	1	10⁻¹	10⁻²	10⁻³	10⁻⁴	10⁻⁵	10⁻⁶	10⁻⁷	10⁻⁸	10⁻⁹
Physical Laws	**Macroscopic**						**Mesoscopic**	**Microscopic**		
Science	Physics (Classical) Biology (Convectional) Engineering (Almost All)				Physics (Solid state) Biology (Micro-bio) Material Science			Physics (Molecular) Biology (Molecular) Chemistry		
Technology	Bulk Technology						Microtechnology	Nanotech		
How to see	Human eyes						Optical Microscope	Electron Microscope	SPM	
Simulation Approach	Macroscale						Mesoscale	Nanoscale	QM	
Successful Model	⟵———— Empirical First Principles ————⟶									

6.2   THE RELATIONSHIP BETWEEN NANOSCIENCE AND MECHANIC QUANTUM

The nanoscale is not just another step toward miniaturization, but a qualitatively new scale. At these sizes, nano systems can exhibit interesting and useful physical behaviors based on quantum phenomena. The new behavior is dominated by quantum mechanics, material confinement in small structures, large interfacial volume fraction, and other unique properties, phenomena and processes. Atom (element)-based chemistry discipline" before the advent of quantum mechanics and electronic theory, Dalton's atom/molecular theory is:

1. Each element consists of picoscale particles called atoms.
2. The atoms of a given element are identical; the atoms of different elements are different in some fundamental way(s).
3. Chemical compounds are formed when atoms of different elements combine with each other. A given compound always has the same relative number in types of atoms.

4. Chemical reactions involve reorganization of atoms (i.e., changes in the way they are bound).

Critical parameters that allowed this important progress evolved around discrete, reproducible features exhibited by each atomic element such as well-defined (a) atomic masses, (b) reactivities, (c) valency, (d) stoichiometries, (e) mass-combining ratios, and (f) bonding directionalities. These intrinsic elemental properties, inherent in all atom-based elemental structures.

Isaac Newton created, more than 300 years ago, classical mechanics by finding the laws of motion for solids and of gravitation between masses. This theory was so successful for the deterministic description of motions. At the beginning of the twentieth century, then, experimental results accumulated which contributed essentially to the emergence of a new physics, quantum physics. Also known as quantum or wave mechanics, this branch of physics was created by Max Planck who showed that the exchange of energy between matter and radiation occurred in discontinuous quantities (quanta). The quantum mechanics is presented as one of the most important and successful theories to solve physical problems. This is totally in the sense of most physicists, who applied, until the 1970s of the twentieth century, in a first quantum revolution quantum mechanics with overwhelming success not only to atom and particle physics but also to nearly all other science branches as chemistry, solid state physics, biology or astrophysics. Because of the success in answering essential questions in these fields fundamental open problems concerning the theory itself were approached only in rare cases. This situation has changed since the last decade of the twentieth century [7].

The "second quantum revolution" as this continuing further development of quantum physical thinking is called by Alain Aspect, one of the pioneers in this field one expects a deeper understanding of quantum physics itself but also applications in engineering. There is already the term "quantum engineering" which describes scientific activities to apply particle wave duality or entanglement for practical purposes, for example, nano-machines, quantum computers, etc. [8, 9] (Fig. 6.1).

The nano world is part of our world, but in order to understand this, concepts other than the normal ones, such as force, speed, weight, etc., must be taken into consideration. The nano world is subject to the laws of quantum physics, yet evolution has conditioned us to adapt to this ever-changing world. This observation has led to further investigate theories based on the laws of physics that deal with macroscopic phenomena. In the macro world, sizes are continuous; however, this is not the case in the nano world. When we investigate and try to understand what is happening on this scale, the way we look at things must be changed. New concepts of quantum physics can only come directly from our surroundings. However, our world is fundamentally quantum. Our common sense in this world has no value in the nano world [10].

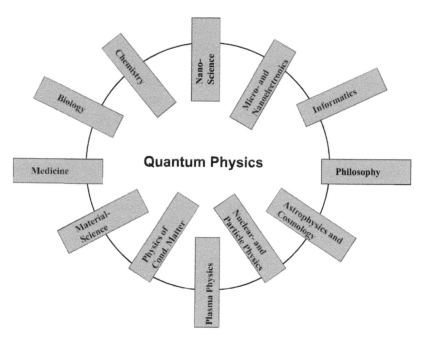

FIGURE 6.1 The relationship between important science branches and the field of quantum physics.

Quantum physics gives a completely different version of the world on the nano metric scale than that given by traditional physics. A molecule is described by a cloud of probability with the presence of electrons at discrete energy levels. This can only be represented as a simulation. All measurable sizes are subject to the laws of quantum physics, which condition every organism in the world, from the atom to the different states of matter. The nano world must therefore be addressed with quantum concepts. Chemistry is quantum. The chemistry of living organisms is quantum. Is the functioning of our brain closer to the concept of a quantum computer or to the most sophisticated microprocessors? All properties of matter are explicable only by quantum physics [10].

Traditional physics, which is certainly efficient and sufficient in the macroscopic domain only deals with large objects (remember that there are nearly 10^{23} atoms per cm^3 in a solid), while quantum physics only deals with small discrete objects. However, the evolution of techniques and the use of larger and larger objects stemming from scientific discoveries make us aware of the quantum nature of the world in all its domains. Everything starts with the atom, the building block of the nano world, and also of our world. In mechanic quantum view, Particles can behave like waves. This property, particularly for electrons, is used in different investigation. On the other hand, waves can also act like particles: the photoelectric effect shows the corpuscular properties of light [10].

6.2.1 THE WAVE FUNCTION AND ITS INTERPRETATION

It has been proven that light waves propagating in space as well as atomic and subatomic particles as electrons moving from one to another spot have one thing in common: Their propagation obeys the laws of wave expansion. In fact, it can be said that everything, matter and energy fields, are simultaneously wave and particle. The correspondence between particle and wave can be expressed by the following relations:

$$E = \frac{1}{2}mv^2 = \hbar\omega \tag{1}$$

$$p = mv = \hbar k = \hbar\frac{2\pi}{\lambda}\frac{k}{|k|} \tag{2}$$

The propagation of a particle, for example, of an electron is described by a wave function. In the simplest case of motion along a straight line a plane wave describes the propagation of the particle, where wave vector and frequency are connected to the particle:

$$\psi(r,t) = ce^{i(k.r-\omega t)} \tag{3}$$

The wave function is a quantity, which is analogous to the wave amplitude of a light field. Its absolute square is identified with an observed intensity after collecting a huge number of electrons on a screen. In particular, the interference pattern in a double slit experiment with electrons is obtained by superimposing two waves originating from two slits at the positions on a remote screen (Fig. 6.2). At a long distance from the source both spherical and cylinder waves (circular holes or slits) can be approximated by plane waves. At the observation point on the remote screen, the superposition of the two wave functions thus yields.

$$\psi = \psi_1 + \psi_2 \qquad \text{with} \qquad \psi_i = c e^{i[k.(r-r_i)-\omega t]} \tag{4}$$

The intensity can be shown as:

$$I = |\psi(r,t)|^2 = |\psi_1|^2 + |\psi_2|^2 + 2c^2 \cos k.(r_2 - r_1) \tag{5}$$

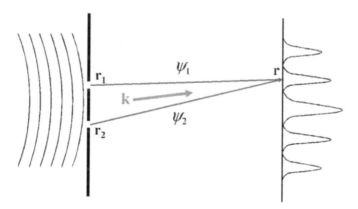

FIGURE 6.2   Scheme of double slit interference of two particle waves.

The wave function is a statistical quantity, which describes only ensemble properties. The probability to find an electron in a volume element is proportional to the volume and, of course, to the probability that is:

$$dP \propto |\psi(r,t)|^2 d^3 r \tag{6}$$

The total probability to find the particle somewhere in the volume must be written as:

$$P(particle \quad in \quad V) = \int_{v} d^3r \, |\psi(r,t)|^2 = 1 \tag{7}$$

The wave function as a probability density must be normalized, in the sense of Eq. (7) over the volume of the whole system considered. Depending on the particular problem the considered volume might be the whole universe.

6.2.2 WAVE PACKET AND PARTICLE VELOCITY

The energy-frequency relation (Eq. (1)), the connection between particle momentum and wave number (Eq. (2)) as well as the description of particle propagation by a wave function and its statistical interpretation (Eqs. (3), (4) and (6)), are the starting point for the formal description of the particle-wave duality.

For a spatially extended wave in the extreme limit, over the whole space-the velocity of a particle cannot be described. The term velocity contains inherently the movement of a particle, an entity, which is more or less limited in its spatial extension.

A particle with a spatial extension in one dimension might be described in simple approximation by a wave function having Gaussian shape.

$$\psi(x) = \frac{1}{\sqrt{2\pi}} \int_{-\infty}^{\infty} a(k) e^{ikx} dk \tag{8}$$

$$a(k) = \frac{1}{\sqrt{2\pi}} \int_{-\infty}^{\infty} \psi(x) e^{ikx} dx \tag{9}$$

$$\psi(x) = \left[2\pi (\Delta x)^2 \right]^{-\frac{1}{4}} \exp\left(-\frac{x^2}{4(\Delta x)^2} \right) \tag{10}$$

$$a(k) = \frac{1}{\sqrt{2\pi}} \int_{-\infty}^{\infty} \left[2\pi (\Delta x)^2 \right]^{-\frac{1}{4}} \exp\left(-\frac{x^2}{4(\Delta x)^2} \right) \exp(-ikx)$$

$$= \frac{1}{\sqrt[4]{(2\pi)^3 (\Delta x)^2}} \int_{-\infty}^{\infty} \exp\left(-\frac{x^2}{4(\Delta x)^2}\right) \exp(-ikx)dx \qquad (11)$$

Finally it can be obtained:

$$a(k) = \left(\frac{2}{\pi}\right)^{\frac{1}{4}} (\Delta x)^{\frac{1}{2}} \exp\left[-(\Delta x)^2 k^2\right] = \left[\frac{4(\Delta x)^2}{2\pi}\right]^{\frac{1}{4}} \exp\left[-(\Delta x)^2 k^2\right] \qquad (12)$$

If Eq. (12) is compared with the common representation of a Gauss distribution as function of k with width Δk, it can be expressed as:

$$a(k) = \frac{1}{2\pi(\Delta k)^2} \exp\left[-\frac{k^2}{4(\Delta k)^2}\right] \qquad (13)$$

The following relation between spatial width Δx of the wave packet and the spread or width of the corresponding wave vector distribution Δk can be obtained as:

$$\Delta k \Delta x = \frac{1}{2} \qquad (14)$$

6.2.3 THE UNCERTAINTY PRINCIPLE

From the representation of a particle by means of a wave packet, we conclude directly that the width Δk of the distribution of wave vectors a(k) that constitute the wave packet is inversely proportional to the spread, that is, the spatial extension of the wave packet (Eq. (14)). For a Gaussian wave packet we quantitatively obtain the relation (Fig. 6.3).

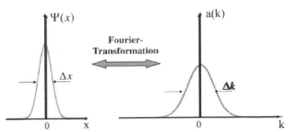

FIGURE 6.3 Gaussian wave packet with a spatial extension (full width at half maximum) and its Fourier representation in the wave number space.

Because of the general rules of Fourier transformation a relation similar to Eq. (14) is always valid:

$$\Delta x.\Delta k \approx 1 \tag{15}$$

This relation between the spatial width Δx of a wave packet and the spread Δk of its Fourier transform leads to an important, typically quantum physical phenomenon.

6.3 THE VISION FOR NANOMATERIALS TECHNOLOGY

Nanomaterials will deliver new functionality and options of material types. A diverse range of nano material building blocks with well-defined properties and stable compositions will enable the design of nanomaterials that provide levels of functionality and performance which are not available in conventional materials.

Manufacturers will combine the benefits of traditional materials and nanomaterials to create new generations of nano material-enhanced products that can be seamlessly integrated into complex systems. In some occasions, nanomaterials will serve as stand-alone devices, providing incomparable functionality. Nanomaterials show a prodigious opportunity for industry to introduce a host of new products that would energize the economy, solve major societal problems, revitalize existing industries, and create entirely new businesses. The race to research, develop, and commercialize nanomaterials is obviously global.

The nano science concept proposed the following: (a) creation of a nanomaterials roadmap focused solely on well defined (i.e., >90% mono-disperse), (0-D) and (1-D) nanoscale materials; (b) these well defined materials were divided into hard and soft nanoparticles, broadly following compositional/architectural criteria for traditional inorganic and organic materials; (c) a preliminary table of hard and soft nanoelement categories consisting of six hard matter and six soft matter particles was proposed. Elemental category selections were based on "atom mimicry" features and the ability to chemically combine or self assemble like atoms; (d) these hard and soft nanoelement categories produce a wide range of stoichiometric nanostructures by chemical bonding or nonbonding assembly. An abundance of literature examples provides the basis for a combinatorial library of hard-hard, hard-soft and soft-soft nano-compounds, many of which have already been characterized and reported. However, many such predicted constructs remain to be synthesized and characterized; (e) based on the presumed conservation of critical module design parameters, many new emerging nano-periodic property patterns have been reported in the literature for both the hard and soft nanoelement categories and their compounds [11] (Fig. 6.4).

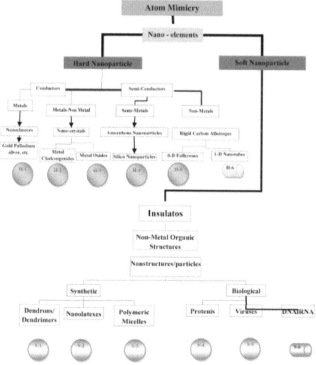

FIGURE 6.4   Nanomaterials classification roadmap.

6.4 THE FUNDAMENTAL IMPORTANCE OF SIZE

Some of the technologies deal with systems on the micrometer range and not on the nanometer range (1–100 nm). In fact, distance scales used in science go to much smaller than nanometers and much larger than meters. All the experience in the macroscopic world suggests that matter is continuous. This, however, leads to a paradox because if matter were a continuum, it could be cut into smaller and smaller pieces without end. If one were able to keep cutting a piece of matter in two, each of those pieces into two, and so on ad infinitum, one could, at least in principle, cut it out of existence into pieces of nothing that could not be reassembled.

Nowadays it can be studied pieces of matter of smaller and smaller size right down to the atom. The important result is that the properties of the pieces start to change at sizes much bigger than a single atom. When the size of the material crosses into the nanoworld, its fundamental properties start to change and become dependent on the size of the piece. It is an important issue to know how the behavior of a piece of material can become critically dependent on its size [12].

Nanomaterials have an increased surface-to-volume ratio compared to bulk materials. Beginning with the most clearly defined category, zero-dimensional nanomaterials are materials wherein all the dimensions are measured within the nanoscale. On the other hand, 1-D nanomaterials differ from 0-D nanomaterials in that the former have one dimension that is outside the nanoscale. This difference in material dimensions leads to needle like shaped nanomaterials. Two-dimensional nanomaterials are more difficult to classify. Three-dimensional nanomaterials, also known as bulk nanomaterials, are relatively difficult to classify. Nowadays it can be studied pieces of matter of smaller and smaller size right down to the atom. The important result is that the properties of the pieces start to change at sizes much bigger than a single atom. Top down approach refers to slicing or successive cutting of a bulk material to get nano sized particle which enables to control the manufacture of smaller, more complex objects, as illustrated by micro and nano electronics. Bottom up approach refers to the buildup of a material from the bottom: atom-by-atom, molecule-by-molecule or cluster-by-cluster, which enables to control the manufacture of atoms and molecules, as illustrated by supra molecular chemistry. Both approaches play very important role in modern industry and most likely in nanotechnology as well. Directed and high rat self-assembly is the efficient methods for nanoelements production, which classified into combined top–down and bottom–up nano manufacturing.

6.5 THERMODYNAMICS AND STATISTICAL MECHANICS OF SMALL SYSTEMS

This section is about the important subjects of computational nanotechnology, namely thermodynamics and statistical mechanics and their applications in molecular systems to predict the properties and performances involving nanoscale structures. A scientific and technological revolution has begun in our ability to systematically organize and manipulate matter on a bottom–up fashion starting from atomic level as well as design tools, machinery and energy conversion devices in nanoscale towards the development of nanotechnology. There is also a parallel miniaturization activity to scale down large tools, machinery and energy conversion systems to micro and nanoscales towards the same goals [13, 14].

Principles of thermodynamics and statistical mechanics for macroscopic systems are well defined and mathematical relations between thermodynamic properties and molecular characteristics are derived. The objective here is to introduce the basics of the thermodynamics of small systems and introduce statistical mechanical techniques, which are applicable to small systems. This will help to link the foundation of molecular based study of matter and the basis for nanoscience and technology.

The subject of thermodynamics of small systems was first introduced by Hill in two volumes in 1963 and 1964 [15] to deal with chemical thermodynamics of mixtures, colloidal particles, polymers and macromolecules. Nanothermodynamics, a term which is recently introduced in the literature by Hill [16, 17], is a revisitation of the original work of Hill mentioned above on thermodynamics of small systems.

6.5.1 THERMODYNAMIC SYSTEM IN NANOSCALE

The definition of a thermodynamic system in nanoscale is the same as the macroscopic systems. In thermodynamics, a system is any region completely enclosed within a well-defined boundary. Everything outside the system is then defined as the surroundings. The boundary may be either rigid or movable. It can be impermeable or it can allow heat, work or mass to be transported through it. In any given situation a system may be defined in several ways.

The simplest system in nanoscale may be chosen as a single particle, like an atom or molecule, in a closed space with rigid boundaries. In the absence of chemical reactions, the only processes in which it can participate are transfers of kinetic or potential energy to or from the particle, from or to the walls. The state for this one-particle system is a set of coordinates in a multidimensional space indicating its position and its momentum in various vector directions.

The set of all the thermodynamic properties of a multiparticle system including its temperature, pressure, volume and internal energy is defined as the thermodynamic state of this system. An important aspect of the relationships between thermodynamic properties in a large, macroscopic and also known as extensive system is the question of how many different thermodynamic properties of a given system are independently variable. The number of these represents the smallest number of properties, which must be specified in order to completely determine the entire thermodynamic state of the system.

6.5.2 LAWS OF THERMODYNAMICS IN NANOSYSTEMS

The application of thermodynamics of large and small systems to the prediction of changes in given properties of matter in relation to energy transfers across its boundaries is based on four fundamental axioms, the Zeroth, First, Second, and Third Laws of thermodynamics. The question whether these four axioms are necessary and sufficient for all systems whether small or large, including nanosystems.

1. The Zeroth Law of thermodynamics consists of the establishment of an absolute temperature scale.
2. The First Law of thermodynamics as defined for macroscopic systems in which no nuclear reactions is taking place is simply the law of conservation of energy and conservation of mass. When, due to nuclear reactions, mass and energy are mutually interchangeable, conservation of mass and conservation of energy should be combined into a single conservation law.

$$dE = \delta Q_{in} + \delta W_{in} \tag{16}$$

Transfer of energy through work mode is a visible phenomenon in macroscopic systems. However, it is invisible in a nanosystem, but it occurs as a result of the collective motion of an assembly of particles of the nanosystem resulting in changes in energy levels of its constituting particles. Transfer of energy through heat mode is also an invisible phenomenon, which occurs in atomic and molecular level. It is caused by a change not of the energy levels but of the population of these levels.

1. Lord Kelvin originally proposed the Second Law of thermodynamics in the nineteenth century. He stated that heat always flows from hot to cold. Rudolph Clausius later stated that it was impossible to convert all the energy content of a system completely to work since some heat is always released to the surroundings. Kelvin and Clausius had macro systems in mind where fluctuations from average values are insignificant in large time scales. According to the Second Law of thermodynamics for a closed (controlled mass) system we have [18].

$$dP_s = ds - \frac{\delta Q_{in}}{T_{ext}} \geq 0 \tag{17}$$

2. The Third Law of thermodynamics for large systems, also known as "the Nernst heat theorem," state that the absolute zero temperature is unattainable. Currently, the third law of thermodynamics is stated as a definition: the entropy of a perfect crystal of an element at the absolute zero of temperature is zero. This definition seems to be valid for the small systems as well as the large systems.

Recent developments in nanoscience and nanotechnology have caused a great deal of interest into the extension of thermodynamics and statistical mechanics to small systems consisting of countable particles below the thermodynamic limit. Hence, if we like to extend thermodynamics and statistical mechanics to small systems in order to remain on a firm basis we must go back to its founders and, like them establish new formalism of thermodynamics and statistical mechanics of small systems starting from the safe grounds of mechanics.

Structural characteristics in nanoscale systems are dynamic, not the static equilibrium of macroscopic phases. Coexistence of phases is expected to occur over bands of temperature and pressure, rather than along just sharp points. The pressure in a nanosystem cannot be considered isotropic and must be generally treated as a tensor.

The Gibbs phase rule loses its meaning, and many phase-like forms may occur for nanoscale systems that are unobservable in the macroscopic counterparts of those systems [15, 19].

6.5.3   STATISTICAL MECHANICS OF NANOSYSTEMS

The objective of statistical mechanics is generally to develop predictive tools for computation of properties and local structure of fluids, solids and phase transitions from the knowledge of the nature of molecules comprising the systems as well as intra and intermolecular interactions.

The accuracy of the predictive tools developed through statistical mechanics will depend on two factors. The accuracy of molecular and intermolecular properties and parameters available for the material in mind and the accuracy of the statistical mechanical theory used for such calculations.

In the case of nano (small) scale there is little or no such data available and the molecular theories of matter in nanoscale are in their infancy. With the recent advent of tools to observe study and measure the behavior of matter in nanoscale it is expected that in a near future experimental nanoscale data will become available.

Recent nanotechnology advances, both bottom–up and top–down approaches, have made it possible to envision complex and advanced systems, processes, reactors, storage tanks, machines and other moving systems which include matter in all possible phases and phase transitions. So in the next section several methods of manufacturing nanoscale will be reviewed.

6.6   NANOELEMENT MANUFACTURING

6.6.1   MANUFACTURING AT NANOSCALE DIMENSIONS

The physicochemical properties of nano-sized materials are really unprecedented, exquisite and sometimes even adjustable in contrast to the bulk phase. For instance, quantum confinement phenomena allow semiconductor nanoparticles to sustain a dilating of their band gap energy as the particle size becomes smaller. Thereby it causes the blue-shifts in the optical spectra and a change in their energy density from continuous to discrete energy levels as the transition moves from the bulk to the nanoscale quantum dot state [20–23]. In addition, interesting electrical properties including resonance tunneling and Coulomb blockade effects are observed with metallic and semiconducting nanoparticles, and endohedral fullerenes and carbon nanotubes can be processed to exhibit a tunable band gap of either metallic or semiconducting properties [24, 25]. These very different phenomena are mainly due to larger surface area-to-volume ratio at the nanoscale compared to the bulk. Thus, the surface forces become more important when the nano-sized materials exhibit unique optical or electrical properties. The surface (or molecular) forces can be generally categorized as electro- static, hydration (hydrophobic, and hydrophilic), Vander Waals, capillary forces, and direct chemical interactions [26, 27]. Based on these forces, the synthesis and processing techniques of these interesting nano-sized materials have been well established as capable of producing high-quality mono- disperse nanocrystals of numerous semiconducting and metallic materials, fullerenes of varying properties, single and multiwall carbon nanotubes, conducting polymers, and other nano-sized systems [28]. The next key step in the application of these materials to device fabrication is undoubtedly the formation of subnanoelements into functional and desired nanostructures without mutual aggregation. To achieve the goal of innovative developments in the areas of microelectronic, optoelectronic and photonic devices with unique physical and chemical characteristics of the nano-sized materials, it may be necessary to immobilize these materials on surfaces and/or assemble them into an organized network [27].

Many significant advances in one- to three-dimensional arrangements in nanoscale have been achieved using the 'bottom–up' approach. Unlike typical top–down photolithographic approaches, the bottom–up process offers numerous attractive advantages, including the substantiation of molecular-scale feature sizes, the potential of three-dimensional assembly and an economical mass fabrication process [29]. Self-assembly is one of the few vital techniques available for controlling the orchestration of nanostructures via this bottom–up technology. The self-assembly process is defined as the autonomous organization of components into well-organized structures. It can be characterized by its numerous advantages such as cost-effective, versatile, facile, and the process seeks the thermodynamic minima of a system, resulting in stable and robust structures [30]. As the description suggests, it is a process in which defects are not energetically favored, thus the degree of perfect organization is relatively high. As described earlier, there are various types of interaction forces by which the self-assembly of molecules and nanoparticles can be accomplished [31, 32].

6.6.2   OVERVIEW OF MOLECULAR SELF-ASSEMBLY

Molecular self-assembly is the assembly of molecules without guidance or management from an outside source. Self-assembly can happen spontaneously in nature, for example, in cells such as the self-assembly of the lipid bilayer membrane. It usually results in an increase in internal organization of the system. Many biological systems use self-assembly to assemble various molecules and structures. Imitating these strategies and creating novel molecules with the ability to self-assemble into supra molecular assemblies is an important technique in nanotechnology [33–35].

In self-assembly, the final desired structure is 'encoded' in the shape and properties of the molecules that are used, as compared to traditional techniques, such as lithography, where the desired final structure must be carved out from a larger block of matter [36].

On a molecular scale, the accurate and controlled application of intermolecular forces can lead to new and previously unachievable nanostructures. This is why molecular self-assembly (MSA) is a highly topical and promising field of research in nanotechnology today. With many complex examples all around in nature, MSA is a widely perceived phenomenon that has yet to be completely understood. Biomolecular assemblies are sophisticated and often hard to isolate, making systematic and progressive analyzes of their fundamental science very difficult. What in fact are needed are simpler MSAs, the constituent molecules of which can be readily synthesized by chemists. These molecules would self-assemble into simpler constructs that can be easily assessed with current experimental techniques [37, 38].

Of the diverse approaches possible for Molecular Self-assembly, two strategies have received significant research attention, electrostatic Self-assembly (or layer-by-layer assembly) and "Self-assembled Monolayers (SAMs). Electrostatic self-assembly involves the alternate adsorption of anionic and cationic electrolytes onto a suitable substrate. Typically, only one of these is the active layer while the other enables the composite multilayered film to be bound by electrostatic attraction. The latter strategy of Self-assembled monolayers or SAMs based on constituent molecules, such as thiols and silanes [39, 40].

For SAMs, synthetic chemistry is used only to construct the basic building blocks (that is the constituent molecules), and weaker intermolecular bonds such as Vander Waals bonds are involved in arranging and binding the blocks together into a structure. This weak bonding makes solution, and hence reversible, processing of SAMs (and in general, MSAs) possible. Thus, solution processing and manufacturing of SAMs offer the enviable goal of mass production with the possibility of error correction at any stage of assembly. It is well recognized that, this method could prove to be the most cost-effective way for the semiconductor electronics industry to produce functional nanodevices such as nanowires, nanotransistors, and nanosensors in large numbers [41, 42].

6.6.3 NANOSELF-ASSEMBLY INVESTIGATION

In the previous sections self-assembly was defined as assembly of its building units. All possible entities (atoms, molecules, colloidal particles) that can take part in this process are self-assembly building units. Building units for nanotechnology systems have more structural hierarchies. Nanotechnology systems can be built not only through self-assembly processes but through an external manipulation as well. All these efforts to create nanotechnology systems can be considered as the processes for assembling nanotechnology systems. We will define this as a nano assembly, which can be stated as a "thermodynamic, kinetic, or manipulative assembly of nano assembly building units." Spontaneous assembly of nano assembly building units will be a great route for building nanotechnology systems [43, 44].

However, assembling them, for example, using an atomic force microscope through a one-by-one type of operation with any type of nano assembly building units will also be a great alternative for creating nanotechnology systems. Figure 6.5 (left-hand side) shows that nano assembled systems are assembled from three basic nano assembly building units. They are a self-assembly building unit, a fabrication-building unit, and a reactive building unit. As will be described in the next section with more details, the structures of all three basic nano assembly-building units can be analyzed based on the concept of segmental analysis. In other words, the segmental analysis that was developed for self-assembly building units can be expanded for the two other types of building units. Figure 6.6 explains this. All three basic nano assembly building units can be analyzed with the three fundamental and two additional segments. And all segments from the three basic nano assembly-building units interact through the force balance with any possible combinations. The whole process resembles the self-assembly process. But it now occurs in a "quasi-three dimensional" way, which is to imply that there are three different types of building units instead of just one (as for self-assembly). The concept of force balance is directly applied not only between self-assembly building units, between fabrication building units, or between reactive building units, but between all three different types of building units as well. This gives us an important insight for the third part of this book that there can be great possibilities for building nano assembled systems once the three basic building units are well identified and the relationships between them are well controlled. The roles of the five segments during the assemblies of nano-assembled systems are the same as for the assemblies of self-assembled systems.

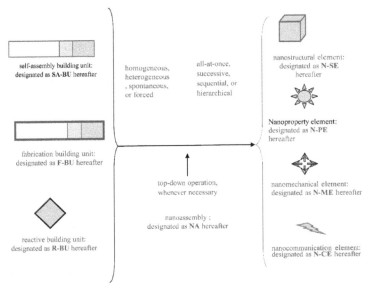

FIGURE 6.5 Three basic nanoassembly building units construct the four nanoelements. Force balance between the nanoassembly building units plays a key role.

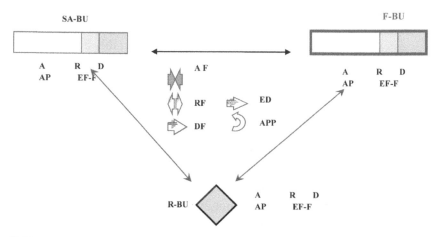

FIGURE 6.6   The fundamental and additional segments of self-assembly building unit (SA-BU), fabrication building unit (F-BU), and reactive building unit (R-BU) can interact through the force balance with any possible combinations. AF, RF, and DF represent attractive force between As, repulsive force between Rs, and directional force between Ds, respectively. A, R, D, AP, and EF-F refer to attractive, repulsive, directional, asymmetric packing, and external force–specific functional segments, respectively. ED is external force–induced directional factor. APP is asymmetric packing process.

Figure 6.5 showed that nanoassembled systems are obtained through nanoassembly with three basic nanoassembly building units. Nanoassembled systems can have a variant range of structures and physical/chemical properties and diverse functional properties. For many nanoassembled systems, these general properties are those that are already known to other existing systems such as macroscopic counter parts. They can be straightly characterized. However, for many others, they can be novel properties that cannot be easily recognized and characterized. There are also nanoassembled systems whose general properties are overlapped by others. The concept of force balance for nanoassembly makes it possible for us to evaluate the specific properties that can be expected from certain nanoassembly building units. It can provide a nice insight when choosing a proper nanoassembly route for a specific nanoassembled system and help clarify intended nanoscale properties (or nanoproperties) with a reasonable degree of accuracy. Four elemental properties (which will be called nanoelements hereafter) for nanoassembled systems are proposed here in order to address these properties in a systematic manner. They are nanostructural element, nanoproperty element, nanomechanical element, and nanocommunication element [45].

The symbols for each nanoelement are also shown in Fig. 6.6. Table 6.2 shows representative examples of the four nanoelements. A nanostructural element is the structural features that are inherited or designed from nanoassembly itself.

TABLE 6.2   Representative Examples of the Four Nanoelements. N-SE, N-PE, N-ME, and N-CE Refer to Nanostructural, Nanoproperty, Nanomechanical, and Nanocommunication Elements, Respectively

	Nanoparticle		**Gating and switching**
N-SE	Nanopore	N-ME	Rotation and oscillation
	Nanofilm		Tweezering and fingering
	Nanotube		Rolling and bearing
	Nanorod		Self-directional movement
	Nano hollow sphere		Capture and release
	Nanofabricated surface		Sensing
N-PE	Surface Plasmon	N-CE	Any macroscale performance by nanointegarated system and energy exchange which are performed by nanomachines
	Quantum size effect		
	Single electron tunneling		
	Surface catalytic activity		
	Mechanical strength		
	Energy conversion		
	Nano-confinement effect		

As shown in the table, most of the nanostructure-based nano assembled systems belong to this. A nano property element is the properties that are inherited, induced, or designed from nano assembly and its framework. Some of them could be the same properties as macroscale counterparts but in the nanoscale while others are those that emerge only when the systems have nanoscale features. A nano mechanical element is the unit operations that are designed to express the motional aspects of nano-assembled systems. Finally, a nano communication element is a signal, energy, or work that is designed to communicate with the macroworld. This nanoelement is almost exclusively for nanofabricated systems, nanointegrated systems, nanodevices, and nanomachines [46].

6.6.4   GENERAL ASSEMBLY DIAGRAM

The outcome of self-assembly is self-assembled aggregate. For nanoassembly, it goes one more step. The apparent initial outcome of nanoassembly is a nanoassembled system. But it is the nanoelements that make nanoassembled systems distinctive from self-assembled aggregates. Self-assembled aggregates have their own characteristic properties, which in many ways are effective, and many applications have been established using them over a wide range of scientific and technological fields. For nanoassembled systems, it is the nanoelements that define their characteristic properties, and with which we are seeking practical applications for nanotechnology systems [47].

Figure 6.7 presents the general rules of nano assembly and their relationship with nanoelements. As a nano assembly becomes more desired (moving toward the right hand direction on the horizontal arrow of attractive interaction-repulsive interaction balance), the nano element that will be expressed is a nano structural element. Typical nano pores, nanoparticles, nano crystals, nano emulsions, and nano composites are more likely to be obtained on this side of the arrow. On the other hand, if a nano assembly moves toward the left-hand side, it is more likely to obtain nano assembled systems that usually need an aid of external force for their assembly. Colloidal crystal is one good example, especially when the size of nano assembly building unit (colloidal particle) is increased. Many top–down operation-based nanoelements are other examples.

When a nano assembly is involved with a directional interaction, the most likely nano element will be a film or surface-based nanoscale operation. Examples include most of the nano-structured films regardless of their detailed morphology. Nano porous film, nano layered film, and nano patterned film are among them. It also includes most of the nanoscale products that are obtained as a result of directional growth (from the spherical-shape) such as nano rods, nano needles, and nanotubes. A good deal of nanofabrication is basically the nanoscale process that is performed on the surface, and thus becomes one prominent example for the upward direction on the vertical arrow. The opposite direction produces nanoelements, too. Some nanoparticles and nano crystals can be obtained at this end. Most of the nano property elements come along with nano structural elements. And they are coupled to each other in many ways [48, 49].

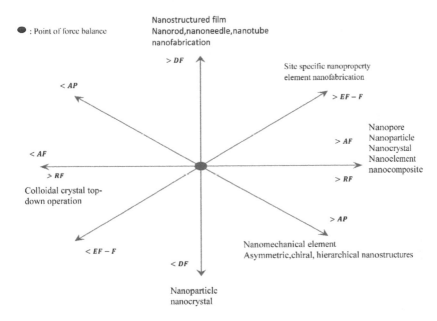

FIGURE 6.7 General rules of nanoassembly and different types of nanoelements. AF, RF, and DF refer to attractive, repulsive, and directional forces, respectively. AP and EF-F are short for asymmetric packing and external force-specific functional segments, respectively.

Most of the nano property elements originate because the nano structural elements are in the nanoscale. And the changes in nano property elements can be feasible because the changes in nano structural elements are practical through nano assembly. The nano assemblies that occur with external force, specific functional and asymmetric packing segments are critical for nano mechanical and nano communication elements. Electron tunneling and Coulomb blockade are good examples. Nanofabrication can take advantage of the unique features of external stimulus-specific nano assembly, too. For a chiral nano assembly, the chirality that is specific on each system can be used for the development of the nanostructures that can take advantage of the uniqueness, which includes highly asymmetric nanostructures, chiral nanoparticles, and some hierarchically constructed multiple-length-scale nanomaterials [50, 51].

It is also important for many unique types of nano mechanical elements. By coupling with the external stimulus-specific nano assembly, the development of nanoelements on this side (right-hand side of both external stimulus-specific and chiral nano assemblies) can be much more fruitful. As far as the application for nanotechnology systems goes, the other side (left-hand side) of both diagonal arrows does not have much use in the development of specific nanoelements [52].

6.6.5   GENERAL TRENDS

Each nanofabricated system is a unique product of each fabrication system. Each nano element of the nanofabricated system is a unique expression of its building units. They can be coupled locally or as a whole. They also can have a synergistic or an antagonistic outcome after the fabrication. All of these aspects have some degree of impact on the nanoelements of the nanofabricated systems. For some cases, different nanofabrication processes become the major reason for differentiating the nanoelements, even though the nanofabricated system might be the same [53].

Figure 6.8 shows a general trend of nanofabrication that covers these aspects from the three approaches. The mass assembling capability of nanofabrication becomes critically important when it goes to industrial scale. Generally, this capability is enhanced where fabrication is performed based on the bottom–up or the bottom–up/top–down hybrid approach. Because of the technical difficulties of top–down techniques and the limitations of the starting bulk materials that are comparable to them, the diversity of building units is much increased when the bottom–up or the hybrid approach is used. More diverse building units mean more diverse nanofabricated systems and more diverse nanoelements that can be explored. Structural diversity, hierarchy, and chirality are also important for widening the practicality of the nanofabricated systems [54].

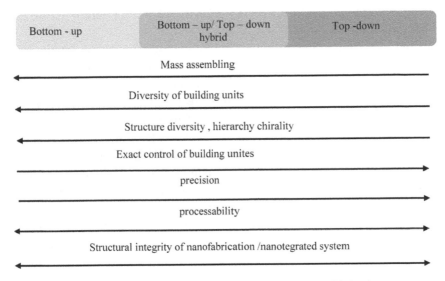

FIGURE 6.8 General trend of the three main approaches to nanofabrication.

It is easier to take advantage of two methods with a bottom–up or hybrid approach. Exact control of the building units is useful when the nano element is determined by the local control of a few main building units. As the top–down approach provides an advantage for this. Generally, the top–down approach has higher precision because of capability of manipulative assembly. Process ability means ease of the fabrication. This factor is important because it determines how practical a specific fabrication can be. It is, however, very much dependent on each nanofabrication system. Another factor is the structural integrity of the nanofabricated systems. It might appear that top–down processed systems would have better structural integrity because they are the products of the bulk materials. But structural integrity is measured not by the absolute strength of the nanofabricated systems but by their relative stability during actual use. As long as they perform the desired functions at given conditions, arguments about their absolute strength are less meaningful. Pure bottom–up fabrication in many cases provides enough, sometimes surprisingly strong, structural strength and resilience for nanofabricated systems to make them function properly even under harsh conditions [55, 56].

6.6.6 INVESTIGATION ONTO SOME NEW STUDIES

The fields of nano science and nanotechnology generally concern the synthesis, fabrication and use of nanoelements and nanostructures at atomic, molecular and supra molecular levels. The nano size of these elements and structures offers significant potential for research and applications across the scientific disciplines, including materials science, physics, chemistry, computer science, engineering and biology. Biological processes and methods, for example, are expected to be developed based entirely on nanoelements and their assembly into nanostructures. Other applications include developing nano devices for use in semiconductors, electronics, photonics, optics, materials and medicine [57].

One class of nanoelements that has garnered considerable interest consists of carbon nanotubes. A carbon nano tube has a diameter on the order of nanometers and can be several micrometers in length. These nanoelements feature concentrically arranged carbon hexagons. Carbon nanotubes can behave as metals or semiconductors depending on their chirality and physical geometry. Other classes of nanoelements include, for example, nano crystals, dendrimers, nanoparticles, nano wires, biological materials, proteins, molecules and organic nanotubes [58].

Although carbon nanotubes have been assembled into different nanostructures, convenient nano tools and fabrication methods to do it have not yet been developed. One obstacle has been the manipulation of individual nanoelements, which is often inefficient and tedious. This problem is particularly challenging when assembling complex nanostructures that require selecting and ordering millions of nanoelements across a large area [59].

To date, nanostructure assembly has focused on dispersing and manipulating nanoelements using atomic force or scanning tunneling microscopic methods. Although these methods are useful for fabricating simple nano devices, neither is practical when selecting and patterning, for example, millions of nanoelements for more complex structures. The development of nano machines or "nano assemblers" which are programed and used to order nanoelements for their assembly holds promise, although there have been few practical advancements with these machines.

The advancement of nanotechnology requires millions of nanoelements to be conveniently selected and simultaneously assembled. Three-dimensional nanostructure assembly also requires that nanoelements be ordered across a large area [60, 61].

Nanoelements have generated much interest due to their potential use in devices requiring nanoscale features such as new electronic devices, sensors, photonic crystals, advanced batteries, and many other applications. The realization of commercial applications, however, depends on developing high-rate and precise assembly techniques to place these elements onto desired locations and surfaces [62].

Different approaches have been used to carry out directed assembly of nanoelements in a desired pattern on a substrate, each approach having different advantages and disadvantages. In electrophoretic assembly, charged nanoelements are driven by an electric field onto a patterned conductor. This method is fast, with assembly typically taking less than a minute; however, it is limited to assembly on a conductive substrate [63]. Directed assembly can also be carried out onto a chemically functionalized surface. However, such assembly is a slow process, requiring up to several hours, because it is diffusion limited. Thus, there remains a need for a method of nano element assembly that is both rapid and not reliant on having either a conductive surface or a chemically functionalized surface [64].

In some cases, devices have a volume element having a larger diameter than the nano element arranged in epitaxial connection to the nano element. The volume element is being doped in order to provide a high charge carrier injection into the nano element and a low access resistance in an electrical connection. The nano element may be upstanding from a semiconductor substrate. A concentric layer of low resistivity material forms on the volume element forms a contact [65].

Semiconductor nano element devices show great promise, potentially outperforming standard electrical, opt-electrical, and sensor- etc. semiconductor devices. These devices can use certain nano element specific properties, 2-D, 1-D, or 0-D quantum confinement, flexibility in axial material variation due to less lattice match restrictions, antenna properties, ballistic transport, wave guiding properties etc. Furthermore, in order to design first rate semiconductor devices from nanoelements, transistors, light emitting diodes, semiconductor lasers, and sensors, and to fabricate efficient contacts, particularly with low access resistance, to such devices, the ability to dope and fabricate doped regions is crucial [66, 67].

As an example the limitations in the commonly used planar technology are related to difficulties in making field effect transistors (FET), with low access resistance, the difficulty to control the threshold voltage in the post-growth process, the presence of short-channel effects as the planar gate length is reduced, and the lack of suitable substrate and lattice-matched hetero structure material for the narrow band gap technologies [68].

One advantage of a nano element FET is the possibility to tailor the band structure along the transport channel using segments of different band gap and or doping levels. This allows for a reduction in both the source-to-gate and gate-to-drain access resistance. These segments may be incorporated directly during the growth, which is not possible in the planar technologies. The doping of nanoelements is challenged by several factors. Physical incorporation of do pants into the nano element crystal may be inhibited, but also the established carrier concentration from a certain do pant concentration may be lowered as compared to the corresponding doped bulk semiconductor. One factor that limits the physical incorporation and solubility of do pants in nanoelements is that the nano element growth temperatures very often are moderate [69].

For vapor-liquid-solid (VLS) grown nanoelements, the solubility and diffusion of do pant in the catalytic particle will influence the do pant incorporation. One related effect, with similar long-term consequences, is the out-diffusion of do-pants in the nano element to surface sites. Though not limited to VLS grown nanoelements, it is enhanced by the high surface to volume ratio of the nano element. Also, the efficiency of the doping, the amount of majority charge carriers established by ionization of donors/acceptor atoms at a certain temperature may be lowered compared to the bulk semiconductor, caused by an increase in donor or acceptor effective activation energy, due to the small dimensions of the nano element. Surface depletion effects, decreasing the volume of the carrier reservoir, will also be increased due to the high surface to volume ratio of the nano element [70].

The above described effects are not intended to establish a complete list, and the magnitudes of these effects vary with nano element material, do pant, and nano element dimensions. They may all be strong enough to severely decrease device performance.

6.7   CONCLUDING REMARKS

The necessary knowledge for obtaining the best nano element manufacturing method is the understanding of principles of nano science, its relationships with mechanic quantum and thermodynamic in nano systems. So these topics were investigated in this chapter with their math relations first. After that, the outcome of self-assembly is reviewed. Self-assembled aggregates have their own characteristic properties, which in many ways are effective, and many applications have been established using them over a wide range of scientific and technological fields. The nano assemblies that occur with external force, specific functional and asymmetric packing segments are critical for nano mechanical and nano communication elements. Electron tunneling and Coulomb blockade are good examples. Each nanofabricated system is a unique product of each fabrication system. Each nano element of the nanofabricated system is a unique expression of its building units.

KEYWORDS

- **Nanoscale Manufacturing**
- **Nanoscience**
- **Quantum**
- **Thermodynamic**

REFERENCES

1. Bainbridge, W. S. (2001). *Societal Implications of Nanoscience and Nanotechnology*. Springer.
2. Shelton, R. D., et al. *Societal Implications of Nanoscience and Nanotechnology.*
3. Park, S. Y. et al. (2008). DNA-Programmable Nanoparticle Crystallization. *Nature, 451(7178)*, 553–556.
4. Islam, M. T. et al. (2005). HPLC Separation of Different Generations of Poly (amidoamine) Dendrimers Modified with Various Terminal Groups. *Analytical Chemistry, 77(7)*, 2063–2070.
5. Islam, M. T., Majoros, I. J., & Baker, J. R. (2005). HPLC Analysis of PAMAM Dendrimer Based Multifunctional Devices. *Journal of Chromatography B, 822(1)*, 21–26.

6. Desai, A., Shi, X., & Baker, J. R. (2008). CE of Poly (amidoamine) Succinamic Acid Dendrimers Using a Poly (vinyl alcohol) Coated Capillary. *Electrophoresis, 29(2),* 510–515.

7. Tegmark, M., & Wheeler, J. A. (2001). *100 Years of the Quantum.* arXiv preprint quant-ph/0101077, *284,* 68–75.

8. Luth, H. (2013). *Quantum Physics in the Nanoworld.* Springer.

9. Aspect, A. (2007). Quantum Mechanics: to be or not to be Local. *Nature, 446(7138),* 866–867.

10. Schodek, D. L., Ferreira, P., & Ashby, M. F. (2009). *Nanomaterials, Nanotechnologies and Design: An Introduction for Engineers and Architects.* Elsevier Science, 560.

11. Tomalia, D. A. (2010). Dendrons/dendrimers: Quantized, Nano-element like Building Blocks for Soft-soft and Soft-hard Nano-Compound Synthesis. *Soft Matter, 6(3),* 456–474.

12. Binns, C. (2010). *Introduction to Nanoscience and Nanotechnology, 14,* John Wiley & Sons, 301.

13. Mansoori, G. A. (2002). *Advances in Atomic & Molecular Nanotechnology.* Nanotechnology, United Nations Tech Monitor, 53.

14. Roco, M. C., Williams, R. S., & Alivisatos, P. (2000). *Nanotechnology Research Directions: IWGN Workshop Report: Vision for Nanotechnology in the Next Decade.* Springer. 316.

15. Hill, T. L. (2004). Thermodynamics of Small Systems. *The Journal of Chemical Physics, 36(12),* 3182–3197.

16. Hill, T. L. (2001). A different Approach to Nanothermodynamics. *Nano Letters, 1(5),* 273–275.

17. Hill, T. L. (2001). Perspective: Nanothermodynamics. *Nano Letters, 1(3),* 111–112.

18. Prigogine, I., & Physicist, C. (1961). *Introduction to Thermodynamics of Irreversible Processes.* Interscience Publishers, New York.

19. Mansoori, G. A. (2002). Organic Nanostructures and Their Phase Transitions. in *Proceedings of the First Conference on Nanotechnology-The Next industrial revolution.*

20. Murphy, C. J. & Coffer, J. L (2002). *Quantum Dots: A Primer.* Applied Spectroscopy, *56(1),* 16A–27A.

21. Carotenuto, G., Pepe, G. P., & Nicolais, L. (2000). *Preparation and Characterization of Nano-sized Ag/PVP Composites for Optical Applications.* The European Physical Journal B-Condensed Matter and Complex Systems, *16(1),* 11–17.

22. Winiarz, J. G. et al. (1999). Observation of the Photorefractive Effect in a Hybrid Organic-inorganic Nanocomposite. *Journal of the American Chemical Society, 121(22),* 5287–5295.

23. Riley, D. J. (2002). *Electrochemistry in Nanoparticle Science.* Current Opinion in Colloid and Interface Science, *7(3),* 186–192.

24. Cleuziou, J. P. et al. (2011). *Electrical Detection of Individual Magnetic Nanoparticles Encapsulated in Carbon Nanotubes.* ACS nano, *5(3),* 2348–2355.

25. Matsui, I. (2005). Nanoparticles for Electronic Device Applications: A Brief Review. *Journal of Chemical Engineering of Japan, 38(8),* 535–546.

26. Mozafari, M., & Moztarzadeh, F. (2011). *Microstructural and Optical Properties of Spherical Lead Sulphide Quantum Dots-based Optical Sensors.* Micro & Nano Letters, IET, *6(3),* 161–164.

27. Kim, S. (2007). Directed Molecular Self-Assembly: Its Applications to Potential Electronic Materials. *Electronic Materials Letters, 3(3),* 109–114.
28. Li, Y., Lu, D., & Wong, C. P. (2009). *Electrical Conductive Adhesives with Nanotechnologies.* Springer, 433.
29. Brust, M., & Kiely, C. J. (2002). *Some Recent Advances in Nanostructure Preparation from Gold and Silver Particles: A Short Topical Review.* Colloids and Surfaces A: Physicochemical and Engineering Aspects, *202(2),* 175–186.
30. Whitesides, G. M., & Grzybowski, B. (2002). Self-assembly at All Scales. *Science, 295(5564),* 2418–2421.
31. Claessens, C. G., & Stoddart, J. F. (1997). π–π Intractions in Self-Assembly. *Journal of Physical Organic Chemistry, 10(5),* 254–272.
32. Zimmerman, S. C. et al. (1996). Self-assembling Dendrimers. *Science, 271(5252),* 1095–1098.
33. Ferreira, M., Cheung, J. H., & Rubner, M. F. (1994). Molecular Self-assembly of Conjugated Polyions: A New Process for Fabricating Multilayer Thin Film Heterostructures. *Thin Solid Films, 244(1),* 806–809.
34. ChEng., J. H., Fou, A. F., & Rubner, M. F. (1994). *Molecular Self-assembly of Conducting Polymers.* Thin Solid Films, *244(1),* 985–989.
35. Antonietti, M., & Förster, S. (2003). Vesicles and Liposomes: A Self-Assembly Principle Beyond Lipids. *Advanced Materials, 15(16),* 1323–1333.
36. Bianchi, E. et al. (2007). Fully Solvable Equilibrium Self-assembly Process: Fine-tuning the Clusters Size and the Connectivity in Patchy Particle Systems. *Journal of Physical Chemistry B, 111(40),* 11765–11769.
37. Zhang, S. (2002). Emerging Biological Materials through Molecular Self-assembly. *Biotechnology Advances, 20(5),* 321–339.
38. Whitesides, G. M., & Boncheva, M. (2002). *Beyond Molecules: Self-assembly of Mesoscopic and Macroscopic Components.* Proceedings of the National Academy of Sciences, *99(8),* 4769–4774.
39. Huc, I., & Lehn, J. M. (1997). *Virtual Combinatorial Libraries: Dynamic Generation of Molecular and Supramolecular Diversity by Self-assembly.* Proceedings of the National Academy of Sciences, *94(6),* 2106–2110.
40. Sukhishvili, S. A. (2005). *Responsive Polymer Films and Capsules via Layer-by-layer Assembly.* Current Opinion in Colloid & Interface Science, *10(1),* 37–44.
41. Ulman, A. (1996). *Formation and Structure of Self-assembled Monolayers.* Chemical Reviews, *96(4),* 1533–1554.
42. Kumar, A., Biebuyck, H. A., & Whitesides, G. M. (1994). Patterning Self-assembled Monolayers: Applications in Materials Science. *Langmuir, 10(5),* 1498–1511.
43. Decher, G. (1997). Fuzzy Nanoassemblies: Toward Layered Polymeric Multicomposites. *Science, 277(5330),* 1232–1237.
44. Ai, H. et al. (2002). Electrostatic Layer-by-layer Nanoassembly on Biological Microtemplates: Platelets. *Biomacromolecules, 3(3),* 560–564.
45. Lee, Y. S. (2012). *Nanotechnology Systems,* in *Self-Assembly and Nanotechnology Systems: Design, Characterization, and Applications,* Wiley Online Library. 33–60.
46. Christensen, J. B. & Tomalia, D. A. (2011). *Dendrimers as Quantized Nano-Modules in the Nanotechnology Field,* in *Designing Dendrimers.* John Wiley & Sons, 1–32.

47. Schwartz, M. (2010). *New Materials, Processes, and Methods Technology.* CRC Press, 712.
48. Mirkin, C. A. & Tuominen, M. (2011). *Synthesis, Processing, and Manufacturing of Components, Devices, and Systems,* in *Nanotechnology Research Directions for Societal Needs in (2020).* Springer, 109–158.
49. Öztürk, S., & Akata, B. (2009). Oriented Assembly and Nanofabrication of Zeolite A Monolayers. *Microporous and Macroporous Materials, 126(3),* 228–233.
50. Takeuchi, K., & Tajima, Y. (2001). Nano-integration: An Ingenuity Driven Approach in Nanotechnology. *Riken Review, 38,* 3–6.
51. Ekinci, K. L. & Roukes, M. L. (2005). Nanoelectromechanical Systems. *Review of Scientific Instruments, 76(6),* 061101-061101-12.
52. Badzey, R. L. et al. (2004). A Controllable Nanomechanical Memory Element. *Applied Physics Letters, 85(16),* 3587–3589.
53. Xu, S. et al. (2006). Integrated Plasma-aided Nanofabrication Facility: Operation, Parameters, and Assembly of Quantum Structures and Functional Nanomaterials. *Vacuum, 80(6),* 621–630.
54. Ho, D., Garcia, D., & Ho, C. M. (2006). Nanomanufacturing and Characterization Modalities for Bio-nano-informatics Systems. *Journal of Nanoscience and Nanotechnology, 6(4),* 875–891.
55. Innocenzi, P., Malfatti, L., & Falcaro, P. (2012). Hard X-rays Meet Soft Matter: When Bottom–up and Top–down Get along Well. *Soft Matter, 8(14),* 3722–3729.
56. Lee, Y. S. (2012). *Nanofabricated Systems: Combined to Function,* in *Self-Assembly and Nanotechnology Systems: Design, Characterization, and Applications.* Wiley Online Library, 333–357.
57. Busnaina, A. & Miller, G. P. (2009). *Functionalized Nanosubstrates and Methods for Three-dimensional Nanoelement Selection and Assembly.* Google Patents.
58. Teredesai, P. V. et al. (2000). Pressure-induced Reversible Transformation in Single-wall Carbon Nanotube Bundles Studied by Raman Spectroscopy. *Chemical Physics Letters, 319(3),* 296–302.
59. Liu, Z. et al. (2000). Organizing Single-walled Carbon Nanotubes on Gold Using a Wet Chemical Self-assembling Technique. *Langmuir, 16(8),* 3569–3573.
60. Shin, S. R. et al. (2009). Fullerene Attachment Enhances Performance of a DNA Nanomachine. *Advanced Materials, 21(19),* 1907–1910.
61. Suo, Z., & Lu, W. (2000). Forces that Drive Nanoscale Self-assembly on Solid Surfaces. *Journal of Nanoparticle Research, 2(4),* 333–344.
62. Abramson, A. R. et al. (2004). Fabrication and Characterization of a Nanowire/polymer-based Nanocomposite for a Prototype Thermoelectric Device. *Journal of Microelectromechanical Systems, 13(3),* 505–513.
63. Akabori, M. et al. (2003). In GaAs Nano-pillar Array Formation on Partially Masked InP (111) B by Selective Area Metal–organic Vapor Phase Epitaxial Growth for Two-dimensional Photonic Crystal Application. *Nanotechnology, 14(10),* 1071–1074.
64. Huang, Y. et al. (2001). Directed Assembly of One-dimensional Nanostructures into Functional Networks. *Science, 291(5504),* 630–633.
65. Bockstaller, M. R., Mickiewicz, R. A., & Thomas, E. L. (2005). Block Copolymer Nanocomposites: Perspectives for Tailored Functional Materials. *Advanced Materials, 17(11),* 1331–1349.

66. Voldman, S. (2009). Nano Electrostatic Discharge. *Nanotechnology Magazine*, *3(3)*, 12–15.

67. Vaia, R. A., & Maguire, J. F. (2007). Polymer Nanocomposites with Prescribed Morphology: Going Beyond Nanoparticle-filled Polymers. *Chemistry of Materials*, *19(11)*, 2736–2751.

68. Thelander, C. et al. (2006). Nanowire-based One-dimensional Electronics. *Materials Today*, *9(10)*, 28–35.

69. Simpkins, B. S. et al. (2008). Surface Depletion Effects in Semiconducting Nanowires. *Journal of Applied Physics*, *103(10)*, 104313–104316.

70. Samuelson, L. I., & Ohlsson, B. J. (2010). *Nanostructures and Methods for Manufacturing the Same*. Google Patents.

CHAPTER 7

A COMPREHENSIVE REVIEW ON AROMATIC POLYESTERS OF N-OXYBENZOIC AND PHTHALIC ACID DERIVATIVES

ZINAIDA S. KHASBULATOVA, and GENNADY E. ZAIKOV

CONTENTS

ABSTRACT

The data on aromatic polyesters based on phthalic and n-oxybenzoic acid derivatives have been presented and various methods of synthesis of such polyesters developed by scientists from different countries for last 50 years have been reviewed.

7.1 INTRODUCTION

The important trend of modern chemistry and technology of polymeric materials is the search for the possibilities of producing materials with novel properties based on given combination of known polymers.

One of the most interesting ways in this direction is the creation of block-copolymers macromolecules of which are the "hybrids" of units differing in chemical structure and composition. Thermodynamical incompatibility of blocks results in stable microphase layering in the majority cases what, finally, allows one to combine the properties of various fragments of macromolecules of block-copolymers in an original way.

Depending on diversity of chemical nature of blocks, their length, number and sequence as well as their ability to crystallize one can obtain materials of structure and properties distinguishing from that of initial components. Here are the huge potential possibilities the practical realization of which has already been started. The most evident is the creation of thermoplastic elastomers (TPEs) – high-tonnage polymeric materials synthesized on the base of principle of block-copolymerization: joining of properties of both thermoplastics and elastomers in one material. The great potentials of block-copolymers caused considerable attention to them within the last years.

Nowadays, all the main problems of physics and physic-chemistry of polymers became closely intertwined when studying block-copolymers: the nature of ordering in polymers, the features of phase separation in polymers and the influence of general molecular parameters on it, the stability of phases at exposing to temperature and power impacts, the features of physical and mechanical properties of microphases and the role of their conjugation.

Existing today numerous methods of synthesis of block-copolymers give the possibility to combine unlimited number of various macromolecules, what already allowed people to synthesize multiple block-copolymers. The thermal and mechanical properties and also the stability of industrial block-copolymers vary in broad limits.

The range of operating temperatures and the thermal stability of TPEs have lately been extended owing to the use of solid blocks of high T_{glass} (of polysulfones or polycarbonates, for instance) combined with soft blocks of low T_{glass}. Moreover, the incompatibility of those blocks results in independence of elasticity modulus on temperature in broad temperature range. Applying appropriate selection of chemical nature of blocks one can also improve the other properties.

Some limitations and unresolved issues still exist in areas of synthesis, analysis and characterization of properties and usage of block-copolymers. This is a good stimulus for the intense researches and development of corresponding fields of industry.

The most preferred methods of synthesis of block-copolymers are the three following. The polymerization according to the mechanism of "live" chains with the consecutive addition of monomers has been used in the first one. The second is based on the interaction of two preliminary obtained oligomers with the end functional groups. The third one is the polycondensation of the second block at expense of end group of primarily obtained block of the first monomer. The second and the third methods allow one to use great variety of chemical structures.

So, to produce block-copolymers one can avail numerous reactions allowing one to bind, within the macromolecule, blocks synthesized by means of polycondensation at expense of joining or cycloreversion.

The second method of synthesis of block-copolymers allows one to produce polymers after various combinations of initial compounds, one among which is that monomers able to enter the reaction of condensation are added to oligomers obtained by polycondensation.

Generally, the bifunctional components are used for creating the block-copolymers of $(-AB-)_n$ type. The necessary oligomers could be obtained by means of either condensation reactions or usual polymerization. The end groups of monomer taken in excess are responsible for the chemical nature of process in case of condensation.

Until now, the morphological studies have been performed mainly on block-copolymers containing two chemically different blocks A and B only. One can expect the revelation of quite novel morphological structures for three-block copolymers, including three mutually incompatible units $(ABC)_n$. And there are few references on such polymers.

Russian and foreign scientists remarkably succeeded in both areas of creation of new inflammable, heat-and thermal resistant polycondensation polymers and areas of development of methods of performing polycondensation and studying of the mechanism of reactions grounding the polycondensation processes [1–6].

The reactions of polycondensation are the bases of producing the most important classes of heterochain polymers: polyarylates, polysulfones, polyarylenesterketones, polycarbonates, polyamides and others [7–12].

Non-equilibrium polycondensation can be characterized by a number of advantages among the polycondensation processes. These are the absence of exchange destructive processes, high values of constants of speed of growth of polymeric chain, etc. However, some questions of nonequilibrium polycondensation still remain unanswered: the mechanism and basic laws of formation of copolymers when the possibility of combination of positive properties of two, three or more initial monomers can be realized in high-molecular product.

The simple and complex aromatic polyesters, polysulfones and polyaryleneketones possess the complex of valuable properties such as high physic-mechanical and dielectric properties as well as increased thermal stability.

There are a lot of foreign scientific papers devoted to the synthesis and study of co polysulfones based on oligosulfones and polyarylenesterketones based on oligoketones.

Because of importance of the problem of creation of thermo-stable polymers possessing high flame- and thermal resistance accompanied with the good physic-mechanical properties, the study of the regularities of formation of copolyesters and block-copolyesters based of oligo sulfone ketones, oligosulfones, oligoketones and oligoformals appeared to be promising depending on the constitution of initial compounds, establishing of interrelations between composition, structure and properties of copolymers.

To improve the basic physic-mechanical parameters and abilities to be reused (in particular, to be dissolved), the synthesis of copolyesters and block-copolyesters has been performed through the stage of formation of oligomers with end reaction-able functional groups.

As the result of performed activities, the oligomers of various chemical compositions have been synthesized: oligosulfones, oligoketones, oligo-sulfoneketones, oligoformals, and novel aromatic copolyesters and block-copolyesters have been produced.

Obtained copoly and block-copolyestersulfoneketones, as well as polyarylates based of dichloranhydrides of phthalic acids and chloranhy-dride of 3,5-dibromine-n-oxybenzoic acid and copolyester with groups of terephthaloyl-bis(n-oxybenzoic) acid possess high mechanical and dielec-tric properties, thermal and fire resistance and also the chemical stabil-ity. The regularities of acceptor-catalyst method of polycondensation and high-temperature polycondensation when synthesizing named polymers have been studied and the relations between the composition, structure and properties of polymers obtained have been established. The synthesized here block-copolyesters and copolyesters can find application in various fields of modern industry (automobile, radioelectronic, electrotechnique, avia, electronic, chemical and others) as thermal resistant construction and layered (film) materials.

7.2   AROMATIC POLYESTERS

Aromatic polyesters are polycondensation organic compounds containing complex ester groups, simple ester links and aromatic fragments within their macromolecule in different combination.

Aromatic polyesters (AP) are thermo-stable polymers; they are ther-moplastic products useful for reprocessing into the articles and materials by means of formation methods from solutions and melts.

Mainly, plastics and films are produced from aromatic copolyesters. APs can be used also as lacquers, fibrous binding agents for synthetic pa-per, membranes, hollow fibers, as additions for semiproducts when obtain-ing materials based on other polymers.

Many articles based on APs appear in industry. The world production of APs increased from 38 millions of tons in 2004 to 50 in 2008, or on 32%. The polymers referred to the class of constructional plastics are dis-tinguished among APs.

Until now, the technical progress in many fields of industry, especially in engineering industry, instrument production, was circumfused namely by the use of constructional plastics. Such exploitation properties of polymers as durability, thermal stability, electroisolation, antifriction properties, optical transparency and others determine their usage instead of ferrous and nonferrous metals, alloys, wood, ceramics and glass [13]. 1ton of polymers replace 5–6 tons of ferrous and nonferrous metals and 3–3.5 tons of wood while the economy of labor expenses reaches 800 man-hours per 1ton of polymers. About 50% of all polymers used in engineering industry are consumed in electrotechnique and electronics. The 80% of all the production of electrotechnique and up to 95% of that of engineering industry has been produced with help of polymers.

The use of construction plastics allows one to create principally new technology of creation the details, machine knots and devices what provides for the high economical efficiency. The construction polymers are well used by means of modern methods: casting and extrusion to the articles operating in conditions of sign-altering loads at temperatures 100–200 °C.

The modern chemical industry gave constructional thermoplastic materials with lowered consumption of material and weight of the machines, devices, mechanisms, reduced power capacities and labor-intensity when manufacturing and exploiting, increased stint.

Nowadays, the radioelectronics, electrotechnical, avia, shipbuilding and fields of industry cannot develop successively without using the modern progressive polymers such as polyarylates, polysulfones, polyesterketones and others, which are perspective construction materials. Only Russian industry involves 50 types of plastics including more than 850 labels and various modifications [14]. As a result, the specific weight of products of engineering areas and several other branches of industry produced with help of plastics grew from 32–35% in 1960 till 85–90% in 1990.

The specific weight of construction plastics among the total world production of plastics reached only 5% in 1975 [15]. But plastics of constructional use prevailed in world production of plastics in 1981–1985 [16].

The introduction of polymers has not only positive effect on the state of already existing traditional areas of industry but also determined the technical progress in rocket and atomic industries, aircraft industry, television, restorative surgery and medicine as a whole et cetera. The world production of construction polymers today is more than 10 million tons.

The thermal and heat-resistant constructional plastics take special place among the polymers. The need for such ones arises from fact that the use of traditional polymeric materials of technological assignment is limited by insufficient working thermal resistance, which is usually less than 103–150 °C.

Two classes of polymers have been used for producing high-tonnage thermally resistant plastics: aromatic polyarylates and polyamides [17]. Starting with these and mutually complementing the properties, materials of high operation properties, and thermo-stable plastics of constructional assignment in the first turn, have been produced.

Some widely used and attractive classes of polymers of constructional assignment are considered in the next section.

7.3 AROMATIC POLYESTERS OF *N*-OXYBENZOIC ACID

The *n*-hydroxybenzoic (*n*-oxybenzoic) acid has been extensively used at polymeric synthesis for the improvement of thermal stability of polymers for the last years [18, 19]. The aromatic polyester "ECONOL," homopolymer of poly *n*-oxybenzoic acid, possesses the highest thermal resistant among the all homopolymeric polyethers [20] and attracts attention for the industry.

The poly *n*-oxybenzoic acid is the linear high-crystalline polymer with decomposition temperature in inert environment of 550 °C [20]. Below this point decomposition goes extremely slow. The loss in weight at 460 °C in air is 3% during 1 h of thermal treatment and than number is 1% at 400 °C. The CO, CO_2 and phenol are released when decomposing in vacuum at temperatures 500–565 °C; the coke remnant is of almost polyphenylene-like structure. It is assumed [27], that decomposition starts from breaking of ester bonds. The energy of activation of the process of fission of up to 30% of fly degradation products is 249, 5 kJ/mol. The following mechanism for the destruction of polymer has been suggested:

The high ordering of polymer is kept till temperature 425 °C. The equilibrium value of the heat of melting of polymers of *n*-oxybenzoic acid is found to be 5, 4 kJ/mol [21].

Several methods for the producing the poly n-oxybenzoic acid have been proposed [20, 23–30], which are based of the high-temperature polycondensation, because the phenol hydroxyl possesses low reaction ability. Usually more reactive chlor anhydrides are used. The polycondensation of n-oxybenzoic acid with blocked hydroxyl group (or of corresponding chloranhydride) happens at temperature of 150 °C and greater. The n-acetoxybenzoic acid [23, 24] or the n-oxybenzoic acid mixed with acylation agent Ac_2O [25–27] at presence of standard catalyst (in two stages: in solution at temperature 180–280 °C and in solid phase at temperature 300–400 °C) as well as chloranhydride of n-oxybenzoic acid [34], or the reaction of n-oxybenzoic acid with acidic halogenating agent (for example, $SOCl_2$, PCl_3, PCl_5) [29] have been used at synthesis of high-molecular polymer aiming to increase the reactivity.

Besides the n-oxybenzoic acid, it's replaced in core derivatives (with F, Cl, Br, J, Me, Et, etc., as replacers) can be used as initial materials in polycondensation processes. However, the low-molecular polymer of logarithmic viscosity equal to 0.16 is formed at polycondensation of methyl-replaced n-oxybenzoic aced in presence of triphenylphosphite [37] which is associated to the elapsing of adverse reaction of intramolecular etherification with triphenyl phosphite, resulting in termination of chain. This reaction does not flow when using PCl_3. High heat- and thermo-stable polyesters of n-oxybenzoic acid of given molecular weight containing no edge COOH-groups can be obtained by heating of n-oxybenzoic acid mixed with dialkylcarbonate at 230–400 °C [20].

The high-molecular polymer can be produced from phenyl ester of n-oxybenzoic acid at presence of butyltitanate in perchloroligophenylene at heating in current nitrogen during 4 h at temperatures 170–190 °C and later after 10 h of exposure to 340–360 °C [20]. Usually [32], Ti, Sn, Pb, Bi, Na, K, Zn or their oxides, salts of acetic, chlorohydratic or benzoic acid are used as catalysts of polycondensation processes when producing polyesters. The polycondensation of polymers of oxybenzoic acids and their derivatives has been performed at absence of catalysts [33] in current nitrogen at temperatures 180–250 °C and pressure during 5–6 h either. The temperature of polycondensation can be lowered [34] if it is carried out in polyphosphoric acid or using activating COOH– group of the substance.

Interesting investigation on study of structure characteristics of polyesters and copolyesters of n-oxybenzoic acid have been performed in Refs. [35, 36]. The introduction of links of n-oxybenzoic acid results in high-molecular compounds of ordered packing of chains, similar to the structure of high-temperature hexagonal modification of homopolymer of *n*-oxybenzoic acid [37, 38]. The presence of fragments of *n*-oxybenzoic acid in macromolecular chain of the polymer not only increases the thermal resistance but improves the physic-chemical characteristics of polymeric materials. Obtained polyesters, consisting of monomers of *n*-oxybenzoic acid and *n*-dioxyarylene (of formula HO-Ar-OH, where Ar denotes bisphenylene, bisphenylenoxide, bisphenylensulfone), possess higher breaking impact strength, than articles from industrial polyester [39]. The data on thermogravimetric analysis of homopolymer of *n*-oxybenzoic acid and its polyesters with 4, 4'-dioxybisphenylpropane, tere-or isophthalic acids are presented in [21]. The temperature of 10% loss of mass is 454 °C, 482 °C, and 504 °C for them correspondingly.

Highly durable, chemically inert, thermo-stable aromatic polyesters can be produced by interaction of polymers containing links of *n*-oxybenzoic acid, aromatic dioxy-compounds, for example of hydroquinone and aromatic dicarboxylic acids [40]. Thermo-and chemically resistant polyesters of improved mechanical strength can be obtained by the reaction of *n*-oxybenzoic acid, aromatic dicarboxylic acids, aromatic dioxy-compounds and diaryl carbonates held in solid phase or in high-boiling solvents [41] at temperature 180 °C and lowered pressure, possibly in the presence of catalysts [42]. Some characteristics of aromatic polyesters based on *n*-oxybenzoic acid are gathered in Table 7.1 [20].

As mentioned above, the thermal stability of polymers is closely tied to the manifestation of fire-protection properties. Many factors causing the stability of the materials to the exposure of high temperatures are characteristic for the fire-resistant polymers too.

Thermally stable polyesters can be produced on the basis of *n*-oxybenzoic acid involving stabilizer (triphenylphosphate) introduced on the last stage. The speed of the weight loss decreases two times, after 3 h of exposure to 500 °C polymers loses 0.87% of mass [43].

The high strength high-modular fiber with increased thermo- and fire-resistance can be formed from liquid-crystal copolyester containing 5–95 mol % of *n*-oxybenzoic acid.

The phosphorus is used as fire-resistant addition; in such a case the oxygen index of the fiber reaches 65% [44].

A 40–70 mol% of the compound of formula Ac-*n*-C_6H_4COOH are used to produce complex polyesters of high mechanical strength [45]. The polyesters with improved physic-chemical characteristics can be obtained by single-stage polycondensation of the melt of 30–60 mol% of n-oxybenzoic acid mixed with other ingredients [46].

TABLE 7.1 Physic-mechanical Properties of Aromatic Polyesters Based on n-Oxybenzoic Acid

Composition of polyesters	Heat resistance on vetch, C, °C	Bending strength, MPa	Flex modulus, N/sm²	Breaking strength, MPa	Elasticity modulus when break, N/sm²
n-oxybenzoic acid, terephthalic acid	–	–	–	35.5	35,000
4,4′-bisphenylquinone				157	47,000
n-oxybenzoic acid, isophthalic acid, hydroquinone, 4,4′-benzophenon dicarboxylic acid	141	163	10,270	–	–
n-oxybenzoic acid, isophthalic acid, hydroquinone, bisphenylcarbonate	–	–	493	120.5	–
n-oxybenzoic acid, isophthalic acid, hydroquinone, 4,4′-bis-hydroxydiphenoxide or bisphenol S	130	160	6100	–	–
n-oxybenzoic acid, isophthalic acid, hydroquinone, 3-chlor-n-oxybenzoic acid and bisphenol D	133	232	10,260	–	–

The most used at synthesizing aromatic polyesters are the halogen-replaced anhydrides of dicarboxylic acids and aromatic dioxy-compounds of various constitutions. However, the growing content of halogens in polyesters obtained from mono-, bis- or tetra-replaced terephthalic acid [46] results in lowering of temperatures of glassing, melting as well as of the degree of crystallinity and to some decrease in mechanical strength for halogen-replaced bisphenols. Chlorine-and bromine-containing antipyrenes worsen the thermal resistance of polyesters and are usually used in a company with stabilizers [47, 48]. (Antipyrenes are chemical substances, either inorganic or organic, containing phosphorus, halogen, nitrogen, boron, and metals, which are used for the lowering of the flammability of polymeric materials).

Besides, the usage of halogen-replaced bisphenols and dicarboxylic acids for synthesizing polyesters of lowered flammability considerably increases the cost of the latter. Consequently, the aromatic inhibiting flame additions are necessary for creating thermally and fire resistant polymeric materials. The introduction of such agents in small amounts would not diminish properties of polyesters.

Accounting for the aforesaid, one can assume that the chemical modification of known thermally resistant polyesters, by means of introduction of solider component (halogen-containing n-oxybenzoic acid) can help to solve the problem of increasing of fire safety of polyesters without worsening of their properties. It is obvious [49] that to replace one should intentionally use bromine atoms, which are more effective in conditions of open fire compared to atoms of other halogens.

So, as it comes from the above, the reactive compound involving halogens are widely used for imparting fire-protection properties to aromatic polyesters, as well as oligomer and polymeric antipyrenes.

The liquid-crystal polyesters (LQPs) became quite popular within the last decades. Those differ in their ability to self-arm, possess low coefficient of linear thermal expansion, have extreme size stability, are very chemically stable and almost do not burn.

LQPs can be obtained by means of polycondensation of aromatic oxyacids (n-oxybenzoic one), dicarboxylic acids (iso- and terephthalic ones) and bisphenols (static copolymers) and also by peretherification of polymers and monomers.

Depending on the degree of ordering, LQPs are classified as smectic, nematic and cholesterol ones. Compounds, able to form liquid-crystal state, consist of long flat and quite rigid, in respect to the major axis, molecules.

LQPs can be obtained by several cases:

- Polycondensation of dicarboxylic acids with acetylic derivatives of aromatic oxy-acids (n-oxybenzoic one) and bisphenols.
- Polycondensation of phenyl esters of aromatic oxy-acids (n-oxybenzoic one) and aromatic dicarboxylic acids with bisphenols.
- Polycondensation of dichloranhydrides of aromatic dicarboxylic acids and bisphenols.
- Copolycondensation of dicarboxylic acids, diacetated of bisphenols and/or acetated of aromatic oxy-acids (n-oxybenzoic one) with polyethyleneterephthalate.

Liquid crystal copolyesters have been synthesized [50–56] on the basis of *n*-acetoxybenzoic acid, acetoxybisphenol, terephthalic acid and m-acetoxybenzoic acid by means of polycondensation in melt. All the copolyesters are thermotropic and form nematic phase. The types of LQPs of *n*-oxybenzoic acid are given in Table 7.2.

LQPs can be characterized by high physic-chemical parameters, see Table 7.3.

The fiber possessing high strength properties is formed from the melt of such copolyesters at high temperatures.

The synthesis of totally aromatic thermo-reactive complex copolyesters based on *n*-oxybenzoic acid, terephthalic acid, aromatic diols and alyphatic acids has been performed by means of polycondensation in melt [57–61].

Complex copolyesters are nematic static copolyesters.

The analysis of patents shows that various methods have been proposed for the production of liquid-crystal complex copolyesters [62–64].

TABLE 7.2 Label Assortment of Thermotropic Liquid-Crystal Polyesters of n-Oxybenzoic acids

Company	Country	Trade label	Remarks
1	2	3	4
Dartco Manufacturing Inc.	USA	Xydar	Polyester based on n-oxybenzoic acid, terephthalic acid and п, п-bisphenol
		SRT-300	Unfilled with normal fluidity
		SRT-500	Unfilled with high fluidity
		FSR-315	50% of talca
		MD-25	50% of glass fiber
		FC-110	Glass-filled
		FC-120	Glass-filled
		FC-130	Mineral filler
		RC-210	Mineral filler
		RC-220	Glass-filled

TABLE 7.2 *(Continued)*

Company	Country	Trade label	Remarks
1	2	3	4
LNP Corp.		Thermocomp	Xydar + 150 polytetrafluor-ethylene (antifriction)
RTP Co		FDX-65194	Xydar compositions: glass-filed with finishing and thermal-stabilizing additions, mineral fillers
Celanese Corp.		Vectra	Polyesters based on n-oxy-benzoic acid and naphthalene derivatives
Celanese Corp.		A-130	30% of glass fiber
		B-130	30% of glass fiber
		A-230	30% of carbon fiber
		B-230	30% of glass fiber
		A-540	40% of mineral filler
		A-900	Unfilled
		A-950	Unfilled
Eastman Kodak Co	USA	Vectron	Copolyesters based on n-oxy-benzoic acid and polyethyl-eneterephthalate
		LCC-10108	
		LCC-10109	
BASF	Germany	Ultrax	Re-replaced complex aro-matic polyester
Bayer A.G.	Germany	Ultrax	Aromatic polyester
ICI	Great Britain	Victrex	Polyester based on n-oxyben-zoic acid and acetoxynaph-toic acid
		SRP-1500G	Unfilled
		SRP-1500G-30	27% of glass fiber
		SRP-2300G	Unfilled (special design)
		SRP-2300G-30	27% of glass fiber (special design)

TABLE 7.2　*(Continued)*

Company	Country	Trade label	Remarks
1	2	3	4
Sumimoto kagaku koge K.K	Japan	Ekonol-RE-6000	Polyester based on n-oxy-benzoic acid, isophthalic acid and bisphenol
Japan Elano Co		Ekonol	Fiber
Mitsubishi chemical Co		Ekonol	Aromatic polyester
Unitica K.K.		LC-2000	Polyesters based on n-oxy-benzoic acid and terephthalic acid
		LC-3000	
		LC-6000	

TABLE 7.3　Physic-mechanical Properties of Some Liquid-Crystal Polyesters of n-Oxybenzoic Acid

Property	Xydar			Vectra				
	SRT-300	SRT-350	FSR-315	A-625 (chem. stable)	A-515 (highly fluid)	A-420 (wear resistant)	A-130 (filled with glass fiber)	C-130 (highly thermal resistant)
Density, g/cm³	1.35	1.35	1.4	1.54	1.48	1.88	1.57	1.57
Tensile limit, MPa	115.8	125.5	81.4	170	180	140	200	165
Modulus in tension, GPa	9.65	8.27	8.96	10	12	20	17	16
Transverse strength, MPa	131.0	131.0	111.7	—	—	—	—	—
Izod impact, J/m								
of samples with cut	128.0	208	75	130	370	100	135	120
of samples without cut	390.0	186	272	—	—	—	—	—
Tensile elongation, %	4.9	4.8	3.3	6.9	4.4	1.3	2.2	1.9

TABLE 7.3 *(Continued)*

Property	Xydar			Vectra				
	SRT-300	SRT-350	FSR-315	A-625 (chem. stable)	A-515 (highly fluid)	A-420 (wear resistant)	A-130 (filled with glass fiber)	C-130 (highly thermal resistant)
Deformation heat resistance (at 1,8 MPa), °C	—	—	—	185	188	225	230	240
Heat resistance on vetch, °C	366	358	353	—	—	—	—	—
Arc resistance, sec	138	138	—	—	—	—	—	—
Dielectric constant at 10^{6}°Hz	3.94	3.94	—	—	—	—	—	—
Dielectric loss tangent at 10^{6}°Hz	0.039	0.039	—	—	—	—	—	—

The description of methods for production of copolymers (having repetitive links from derivatives of *n*-oxybenzoic acid, 6-oxy-2-naphthoic acid, terephthalic acid and aromatic diol) formable from the melt is reported in papers [65–68]. Each repetitive link is in certain amounts within the polymer. All these copolymers are used as protective covers.

The thermotropic liquid-crystal copolyesters can be synthesized also on the basis of *n*-oxybenzoic and 2, 6-oxynaphthoic acid at presence of catalysts (sodium and calcium acetates) [69, 70]. The catalyst sodium acetate accelerates the process of synthesis. Calcium acetate accelerates the process only at high concentration of the catalyst and influences on the morphology of complex copolyesters.

The syntheses of copolymers of *n*-oxybenzoic acid with polyethyleneterephthalate and other components are possible too [71–75]. It is established that copolymers have two-phase nature: if polyethyleneterephthalate is introduced into the reaction mix then copolymers of block-structure are formed.

The properties of liquid-crystal copolyesters based on *n*-oxybenzoic acid and oxynaphthoic acid (in various ratios and temperatures) are described in papers [76–81]. It was shown therein that the plates from such polymers formed by die-casting are highly anisotropic which results in appearance of the layered structure. Those nematic liquid crystals had ferroelectric ordering. Following substances are characterized in [82–85]: aromatic copolyesters of *n*-oxybenzoate/bisphenol A; those based on *n*-oxybenzoic acid, bisphenol and terephthalic acid; based on 40 molar % *n*-oxybenzoic acid, 30 molar % *n*-hydroquinone and isophthalic acid; and those based on *n*-oxybenzoic acid, *n*-hydroquinone and 2,6-naphthalenedicarboxylic acid. The influence of temperature, warm-up time and heating rate on the properties of copolyesters was studied: the glass-transition temperature was shown to increase with warm-up expanded. The number of works is devoted to the study of the complex of physic-chemical properties of liquid-crystal copolyesters based on *n*-oxybenzoic acid and polyethyleneterephthalate [86–97]. It was established that mixes up to 75% of liquid-crystal compound melt and solidify like pure polyethyleneterephthalate. Copolyesters, containing less than 30% of *n*-oxybenzoic acid, are in isotropic glassy state while copolyesters of higher concentration of the second compound are in liquid-crystal phase and can be characterized by bigger electric inductivity, than in glassy state. This distinction in caused by the existence of differing orientational distribution of major axes in relation to the direction of electric field in various structure instances of copolyesters. The data of IR-spectroscopy reveal that components of the mix interact in melt by means of reetherification reaction. Increasing pressure, decreasing free volume and mobility in the mix, one can delay the reaction between the compounds.

The investigation of the properties of liquid-crystal copolyesters based on *n*-oxybenzoic acid/polyethyleneterephthalate and their mixes with isotactic polypropylene (PP), polymethylmethacrylate, polysulfone, polyethylene-2, 6-naphthalate, copolyester of *n*-oxybenzoic acid 6,2-oxynaphthoic acid continues in papers [98–106]. Specific volume, thermal expansion coefficient α and compressibility β were measured for PP mixed with liquid-crystal copolymer *n*-oxybenzoic acid/polyethyleneterephthalate. The high pressure results in appearance of ordering in melted PP. The increase of α and β for all mixes studied (100–25% PP) has been observed at temperatures about melting point of PP having those parameters for liquid-crystal copolyester changed inconsiderably. Increasing content of liquid-crystal copolyester in the mix remarkably decreases the value of α both in solid state and in area of melting PP. The latter has to be accounted for when producing articles. The liquid-crystal compound is grouped in shape of concentric cylinders of various radii in mixes with polysulphones. It becomes responsible for the viscous properties of the mixes at above conditions.

Piezoelectric can be produced from liquid-crystal polymers based on n-oxybenzoic acid/polyethyleneterephthalate and n-oxybenzoic acid/oxynaphthoic acid. Time-stability and temperature range of efficiency of piezoelectric from polymer made of 6.2-oxynaphthoic acid are higher compared to that of polyethyleneterephthalate.

It was found that the use of 30% liquid-crystal copolyester of n-oxybenzoic acid/polyethyleneterephthalate makes polymethylmethacrylate 30% more durable and increases the elasticity modulus on 110%, with the reprocessibility being the same.

The structure and propertied of various liquid-crystal copolymers are studied in papers [107–112]. For example, copolymers based on n-oxybenzoic acid, polyethyleneterephthalate, hydroquinone and terephthalic acid, are studied in [107] the introduction of the mix hydroquinone/terephthalic acid accelerates the process of crystallization and increases the degree of crystallinity of polyesters.

The existence of two structure areas of melts of copolyesters, namely of low-temperature one (where high-melting crystals are present in nematic phase) and high-temperature other (where the homogeneous nematic alloy is formed) is revealed when studying the curves of fluxes of homogeneous and heterogeneous melts of copolyesters based on polyethyleneterephthalate and acetoxybenzoic acid [108].

The curves of flux of liquid-crystal melt are typical for viscoplastic systems while the trend for the flow limit to exist becomes more evident with increasing molar mass and decreasing temperature. Essentially higher quantities of molecular orientation and strength correspond to extrudates obtained from the homogeneous melt compared to extrudates produced from heterophase melt. The influence of high-melting crystallites on the process of disorientation of the structure and on the worsening of durability of extrudates becomes stronger with increasing contribution of mesogenic fragment within the chain.

When studying structure of liquid-crystal copolymers based on n-oxybenzoic acid/polyethyleneterephthalate and m-acetoxybenzoic acid by means of IR-spectroscopy, ^{1}H nuclear magnetic resonance and large angle X-ray scattering, the degree of ordering in copolymers was shown [109] to increase if concentration of n-oxybenzoic units went from 60 till 75%.

When studying properties of copolymers n-oxybenzoate/ ethyleneterephthalate/m-oxybenzoate with help of thermogravimetry [110] the thermo-decomposition of copolymers was found to happen at temperatures 450–457°C in N_2 and 441–447°C in air. The influence of the ratio of n-and m-isomers was regarded, the coal yield at $T > 500$ °C found to be 42–6% and increasing with growing number of n-oxybenzoic units.

The mixing of melts poly 4, 4'-oxybenzoic acid/polyethyleneterephthalate was studied in Ref. [111]: the kinetic characteristics of the mix became incompatible at reetherification.

The measurement of glassing temperature T_{glass} [112] of thermotropic liquid-crystal polyesters synthesized from 4-acetoxybenzoic acid (component A), polyethyleneterephthalate (component B) and 4-acetoxy-hydrofluoric acid revealed that the esters could be characterized by two phases, to which two glassing temperatures correspond: T_{glass} = 66–83°C and T_{glass} = 136–140°C. The lower point belongs to the phase enriched in B, while higher one – to phase enriched in A.

Liquid-crystal polymers are most applicable among novel types of plastics nowadays. The chemical industry is considered to be the one of perspective areas of using liquid-crystal polymers where they can be used, because of their high thermal and corrosion stability, for replacement of stainless steel and ceramics.

Electronics and electrotechnique are considered as promising areas of application of liquid-crystal polymers. In electronics, however, liquid-crystal polymers meet acute competition from cheaper epoxide resins and polysulfone. Another possible areas of using high-heat-resistant liquid crystal polymers are avia, space and military technique, fiber optics (cover of optical cable and so on), auto industry and film production.

The question of improving fire-resistance of aromatic polyesters is paid more attention last time. Polymeric materials can be classified on criterion of combustibility: noncombustible, hard-to-burn and combustible. Aromatic polyesters enter the combustible group of polymers self-attenuating when taken out of fire.

The considerable fire-resistance of polymeric materials and also the conservation of their form and sizes are required when polymers are exploited to hard conditions such as presence of open fire, oxygen environment, and exposure to high-temperature heat fluxes.

On the assumption of placed request, the extensive studies on both syntheses of aromatic polyesters of improved fire-resistance and modification of existing samples of polymers of given type have been carried out. The most used methods of combustibility lowering are following:

- Coating by fireproof covers;
- Introduction of filler;
- Directed synthesis of polymers;
- Introduction of antipyrenes;
- Chemical modification.

The chemical modification is the widely used, easily manageable method of improving the fire resistance of polymers. It can be done synthetically, simultaneously with the copolymerization with reactive modifier via, for example, introduction of replaced bisphenols, various acids, and other oxy-compounds. Or it can be done by addition of reactive agents during the process of mechanic-chemical treatment or at the stage of reprocessing of polymer melt [113].

The most acceptable ways of modification of aromatic polyesters aimed to get self-attenuating materials of improved resistance to aggressive media are the condensation of polymers from halogen-containing monomers, the combination of aromatic polyesters with halogen-containing compounds and the use of halogen-involving coupling agents [114, 115].

Very different components (aromatic and aliphatic) can become modifiers. They inhibit the processes of combustion and are able to not only to attach some new features to polymers but also improve their physical properties.

7.4   AROMATIC POLYESTERS OF TEREPHTHALOYL-BIS-(N-OXYBENZOIC) ACID

Among known classes of polymers the aromatic polyesters with rigid groups of terephthaloyl-bis-(n-oxybenzoic) acid in main chain of next formula attract considerable attention:

The polymers with alternating terephthaloyl-bis (*n*-oxybenzoatomic) rigid (R) mesogenic groups and flexible (F) decouplings of various chemical structure within the main chain (RF-copolymers) are able to form liquid-crystal order in the melt. The interest to such polymers is reasoned by fact that dilution of mesogenic "backbone" of macromolecule by flexible decouplings allows one to change the temperature border of polymer transfer from the partially crystalline state into the liquid-crystal one and also to change the interval of existence of liquid-crystal melt. The typical representatives of such class of polymers are polyesters with methylene flexible decouplings.

These polymers have been synthesized by high-temperature polycondensation of terephthaloyl-bis (*n*-oxybenzoylchloride) with appropriate diols in high-boiling dissolvent under the pressure of inert gas. Bisphenyloxide is used as dissolvent.

The features of conformation, orientation order, molecular dynamics of mentioned class of polymers are studied in [116–119] on the basis of polydecamethyleneterephthaloyl-bis(*n*-oxybenzoate), P-10-MTOB, which range of liquid-crystallinity is from 230 °C till 290 °C.

The study of polyesters with terephthaloyl-bis (*n*-oxybenzoate) groups continues in papers [120–123]. The oxyethylene $(CH_2CH_2O)_n$ and oxypropylene $(CH_2CHCH_3O)_n$ groups are used as flexible decouplings. It was found for oxyethylene decouplings that mesophase did not form if flexible segment was three times longer than rigid one. The folding of flexible decoupling was found to be responsible for the formation of ordered mesophase in polymers with long flexible decoupling. It was observed that polymers are able to form smectic and nematic liquid-crystal phases, the diapason of existence of which is determined by the length of flexible fragments.

The reported in Ref. [124] were the data on polymers containing methylene siloxane decouplings in their main chain:

The polymer was produced by heating of dichloranhydride of tere-phthaloyl-bis (*n*-oxybenzoic) acid, 1,1,3,3-tetramethylene-1,3-bis-(3-hydroxypropyl)-bis-siloxane and triethylamine in ratio 1:1:2, respectively, in environment of chloroform in argon atmosphere. The studied were the fibers obtained by mechanical extrusion from liquid-crystal melt when heating initial sample. The polymer's characteristic structure was of smectic type with folding location of molecules within layers.

The study of liquid-crystal state was performed in [125] on polymers with extended (up to 5 phenylene rings) mesogenic group of various lengths of flexible oxyethylene decouplings:

All polymers were obtained by means of high-temperature acceptor-free polycondensation of terephthaloyl-bis (*n*-oxybenzoylchloride) with bis-4-oxybenzoyl derivatives of corresponding polyethylene glycols in solutions of high-boiling dissolvent in current inert gas. Copolyesters with mesogenic groups, extended up to 5 phenylene cycles and up to 15–17 oxyethylene links, are able to form the structure of nematic type.

Polymers can also form the systems of smectic type [126]. The studied in Ref. [127] was the mesomorphic structure of polymer with extended group polyethylene glycol-1000-terephthaloyl-bis-4-oxybenzoyl-bis-4'-oxybenzoyl-bis-4''-oxybenzoate:

This polymer (of formula $R=CH_2CH_2 (OCH_2CH_2)_{18-20}$) was synthesized by means of high-temperature acceptor-free polycondensation of terephthaloyl-bis-4-oxybenzoate with bis-(4-oxybenzoyl-4'-oxybenzoyl) derivative polyethylene glycol-1000 in the environment of bisphenyloxide [128]. The X-ray diffraction pattern revealed that the regularity of layered order weakened with rising temperature while ordering between mesogenic groups in edge direction kept the same. That type of specific liquid-crystal state in named polymer formed because of melting of layers including flexible oxyethylene decouplings at keeping of the ordering in transversal direction between mesogenic groups.

The Ref. [129] was devoted to the study of the molecular mobility of polymer which formed mesophase of smectic type in temperature range 223–298°C. Selectively deuterated polymers were synthesized to study the dynamics of different fragments of polymer in focus. That polymer was found to have several coexisting types of motion of mesogenic fragment and decoupling at the same temperature. There were massive vibration of phenylene cycles with varying amplitude and theoretically predicted movements of polymethylene chains [130]: transgosh-isomerization and translation motions involving many bonds. Such coexistence was determined by the phase microheterogeneity of polymers investigated.

The process of polycondensation of terephthaloyl-bis (*n*-oxybenzoyl-chloride) with decamethyleneglycol resulting in liquid-crystal polyester was studied in [131]. The monomers containing groups of complex esters entered into the reaction with diols at sufficiently high temperature (190 °C).

The relatively low contribution of adverse reactions compared to the main one led to regular structure of final polyester. The totally aromatic polyesters of following chemical constitution were studied in Ref. [132]: polymer "ΦTT-40" consisting of three monomers (terephthaloyl-bis (*n*-oxybenzoic) acid, terephthalic acid, phenylhydroquinone and resorcinol) entering to the polymeric chain in quantities 2:3:5. Terephthalic acid was replaced with monomer containing phenylene cycle in meta-place in polymer "ΦΓP-80."

terephthaloyl-bis(*n*-oxybenzoic) acid (*A*),

terephthalic acid (*B*),

phenylhydroquinone (*C*),

resorcinol (*D*).

The monomers entered to the polymeric chain in ratio $A: C: D$=5:4:1. These polymers melted at temperatures 300–310°C and switched to liquid-crystal state of nematic type. The monomers in "ΦTT-40" were found to be distributed statistically.

The influence of chemical structure of flexible decouplings in polytere-phthaloyl-bis (*n*-oxybenzoates) on their mesogenic properties was studied in Ref. [131]. The studied were the polymers, which had asymmetrical centers, introduced into the flexible methylene decoupling and also complex ester groups:

$$R=CH(CH_3)CH_2CH_2CH_2CH_2 \text{ (I); } R=CH(R')COOCH_2CH_2 \text{(II, I)}$$
$$\text{where } R'=CH(CH_3)_2 \text{ (II) or } CH_2CH_2(CH_3) \text{ (III).}$$

The polymer I was synthesized by means of high-temperature acceptor-free polycondensation from dichloranhydride of terephthaloyl-bis(*n*-oxybenzoic) acid and 2-methylhexa-methylene-1, 5-diol in inert dissolvent (bisphenyloxide at 200 °C). The phase state of polymer I at room temperature was found to depend on the way of sample preparation. The samples obtained from polymer immediately after synthesis and those cooled from melt were in mesomorphic state, but dried from the solutions in trifluoroacetic acid were in partially crystalline state.

The polymers II and III were in partially crystalline state at room temperature irrespective from the method of production.

So, the introduction of asymmetrical center into the flexible pentamethylene decoupling did not change the type of the phase state in the melt and did not influence on the temperature of the transitions into the area of existing of liquid-crystal phase. However, increasing of the rigidity of methylene decouplings via the introduction of complex-ester groups and enlarging of the volume of side branches considerably influenced on the character of intermolecular interaction what resulted in the formation of mesomorphic state of 3D structure during the melting of polymers.

The conformational and optic properties of aromatic copolyesters with links of terephthaloyl-bis (*n*-oxybenzoic) acid, phenylhydroquinone and resorcinol, containing 5% (from total number of para-aromatic cycles) of m-phenylene cycles within the main chain were studied in [134]. The polymer had the structure:

It was determined that the length of statistic Kuhn segment A was 200 ± 20Å, the degree of dormancy of intramolecular rotations $(\sigma^{-2})^{1}/_{2}=1.08$, the interval of molar masses started with 3.1 and ended 29.9×10^{3}.

The analysis of literature data shows that the use of bisphenyl derivatives as elements of structure of polymeric chain allows one to produce liquid crystal polyesters, reprocessible from the melt into the articles of high deformation-strength properties and high heat resistance.

The polyethers containing mesogenic group and various bisphenylene fragments were synthesized in Ref. [135]:

links of 4-oxybisphenyl-4-carbonic acid,　links of n-oxybenzoic acid,

links of terephthalic acid,

links of dihydroxylic bisphenyls of (3,3'-dioxybisphenyl).

All were linked with various decouplings within the main chain of next kinds:

poly(alkyleneterephthaloyl-bis-4-oxybisphenyl-4'-carboxylate)
$[\eta]=0{,}78$ deciliter/gram $T_{glass}=160°C$, $T_{vapor}=272°C$, where $R=(CH_2)_6$; $(CH_2)_{10}$

poly(oxyalkyleneterephthaloyl-bis-4-oxybenzoyl-4′-oxybenzoate
where R = $-CH_2-CH_2OCH_2-CH_2-$; $-(-CH_2CH_2O-)_2-CH_2CH_2-$ oxyethylene decouplings; polyethylene glycols (PEG) PEG 200, PEG 300,
PEG 400, PEG 600, PEG 1000 – polyoxyethylene decouplings of various
molecular weights and copolymers based on 3,3′-dioxybisphenyl with elements of regularity within the main chain:

links of terephthaloyl-bis(n-oxybenzoic) acid – links of 3,3′-dioxybi-
sphenyl

links of terephthalic acid-links of terephthaloyl-bis(n-oxybenzoic) acid-
links of 3,3′-dioxybisphenyl.

The synthesized in Ref. [136] were the block-copolymers containing the links of terephthaloyl-bis (n-oxybenzoate) and polyarylate. It was
found that such polymers had liquid-crystal nematic phase and the biphase
separation occurred at temperatures above 280 °C.

The thermotropic liquid-crystal copolyester of polyethyleneterephthalate and terephthaloyl-bis (n-oxybenzoate) was described in Ref. [137]. The
snapshots of polarized microscopy revealed that copolyester was nematic
liquid-crystal one.

When studying the reaction of reetherification happening in the mix
50:50 of polybutyleneterephthalate and complex polyester, containing mesogenic sections from the remnants of n-oxybenzoic (I) and terephthalic
(II) acids separated by the tetramethylene decouplings the content of four
triads with central link (I) and three triads with central link II was determined in Ref. [138]. The distribution of triad sequences approximated to

the characteristic one for statistical copolymer with increasing of the exposure time.

The relaxation of liquid-crystal thermotropic polymethyleneterephthaloyl-bis(n-oxybenzoate) was studied in Ref. [139]: the β-relaxation occurred at low temperatures while α-relaxation took place at temperatures above 20 °C (β-relaxation is associated with local motion of mesogenic groups while α-relaxation is caused by the glassing in amorphous state). It was established thin the presence of ordering both in crystals and in nematic mesophase expanded relaxation.

So, the liquid-crystal nematic ordering is observed, predominantly, in considered above polymeric systems containing links of terephthaloyl-bis (n-oxybenzoic) acid [140] and combining features of both polymers and liquid crystals. Such ordering can be characterized by the fact that long axes of mesogenic groups are adjusted along some axis and the far translational ordering in distribution of molecules and links is totally absent.

The analysis of references demonstrates that polyesters with links of terephthaloyl-bis (n-oxybenzoic) acid can be characterized by the set of high physic-mechanical and chemical properties.

That is why we have synthesized novel copolymers on the basis of dichloranhydride of terephthaloyl-bis (n-oxybenzoic) acid and various aromatic oligoesters.

7.5   AROMATIC POLYARYLENESTER KETONES

The acceleration of technical progress, the broadening of assortment of chemical production, the increment of the productivity of labor and quality of items from plastics are to a great extent linked to synthesis, mastering and application of new types of polymeric materials. New polymers of improved resistance to air, heat form-stability, and longevity appeared to meet the needs of air and space technique. Filling and reinforcement can help to increase the heat resistance on about 100 °C. Plastics enduring the burden in temperature range 150–180°C are called "perspective" while special plastics efficient at 200 °C are named "exotic."

The most promising heat-resistant plastics for hard challenges are polysulfones.

Today, the widely used aromatic polysulfones as constructional and electroisolating materials are those of general formula $[-O-Ar-SO_2-]_n$ where R stands for aromatic radical.

The modified aromatic polysulfones attract greater interest every day. The introduction of ester group into the chain makes the molecule flexible, elastic, more "fluid" and reprocessible. The introduction of bisphenylsulphonic group brings thermal resistance and form stability. Non-modified polysulfones are hardly reprocessible due to the high temperature of melt and are not used for technical purposes.

Aromatic polysulfones are basically linear amorphous polyerylensulfonoxides. They are constructional thermoplasts, which have sulfonic groups-SO_2- in their main chain along with simple ester bonds, aliphatic and aromatic fragments in different combinations.

There are two general methods known for the production of polyarylenesulfonoxides from hydroxyl-containing compounds [8, 141]. These are the polycondensation of disodium or dipotassium salts of bisphenols with dahalogenbisphenylsulfone and the homopolycondensation of sodium or potassium salts of halogenphenoles. One should notice that mentioned reactions occur according to the bimolecular mechanism of nucleophylic replacement of halogen atom within the aromatic core.

The first notion of aromatic polysulfones as promising constructional material was met in 1965 [142–145]. The most pragmatic interest among aromatic polysulfones attracts the product of polycondensation of 2,2-bis (-4-oxyphenyl) of propane and 4,4'-dichlorbisphenyldisulfone:

The degree of polymerization of industrial polysulfones varies from 60 till 120 what corresponds to the weights from 30,000 to 60,000.

The other bisphenols than bisphenylolpropane (Diane) can be used for synthesizing polysulfones, and the constitution of ingredients make considerable effect on the properties of polymers.

Linear polysulfones based on bisphenylolpropane and containing iso-propylidene groups in the chain are easily reprocessible into the articles and have high hydrolysis stability. The presence of simple ester links in the polymeric chains makes them more flexible and durable. The main effect on properties of such polysulfones is produced by sulfonic bond, which makes the polymer more stable to oxidation and more resistant to heat. Above properties of polysulfones along with the low cost of bisphenylol-propane change them into almost ideal polymers for constructional plastics. Polysulfones of higher heat-resistance can be obtained on the basis of some other bisphenols.

The industrial production of polysulfones based on 2,2-bis-(4oxyphenyl) of propane and 4,4'-dichlorbisphenylsulfone was started by company "Union Carbid" (USA) in 1965 [146].

The manufacture of given aromatic polysulfones was also launched by company "ICI" in Great Britain under the trademark "Udel" in 1966. Various labels of constructional materials (P-1700, P-1700-06, P-1700-13, P-1700-15, etc.) are developed on the basis of this type of polysulfones.

Aromatic polysulfones are soluble in different solvents: good dissolution in chlorized organic solvents and partial dilution in aromatic hydrocarbons.

By its thermo-mechanical properties the aromatic polysulfone on the basis of bisphenylolpropane takes intermediate place between polycarbonate and polyarylate of the same bisphenol. The glassing temperature of the polysulfone lies in the ranges of 190–195 °C, heat resistance on vetch is 185 °C. The given polysulfone is devised to be used at temperatures below 150 °C and is frost-resistant material (–100 °C).

One of the most valuable properties of polysulfones is well creep resistance, especially at high temperatures. The polysulfone's creep deformation is 1.5% at 100 °C and after $3.6 \quad 10^{60}$ sec loading of 21 MPa, which is better than that of others. Their long-term strength at high temperature is also better. Thus, the polysulfone can be used as constructional material instead of metals.

The flow limit of polysulfone is 71.5 MPa, the permanent strain after rupture is 50–100%. At the same time the elasticity modulus (25.2 MPa) points on sufficient rigidity of material, which is comparable to that of polycarbonate. Impact strength on Izod is 7–8 kJ/m^2 at 23 °C with notch.

The strength and durability of polysulfones keep well at high temperatures. This fact opens possibilities to compete with metals in such areas where other thermoplasts are worthless.

The possibilities of using polysulfones in parts of high-precision articles arise from its low shrinkage (0.7%) and low water absorption (0.22% after 5.64×10^4 sec).

Aromatic polysulfone on the basis of bisphenylolpropane is relatively stable to thermo-oxidation destruction, because the sulfur is in its highest valence state in such polymers; electrons of adjacent benzene nuclei shift, under the presence of sulfur, to the side of sulfogroups what causes the resistance to oxidation.

The polysulfone can operate long at temperature up to 140–170 °C, the loss of weight after 2.52×10^7 sec of loading at 125 °C is less than 0.25%; polymer loses 3% of its mass after 3.24×10^7 sec of exposure to 140 °C in oxygen.

The results of thermal destruction of polysulfone in vacuum have revealed that the first product of decomposition at 400 °C was the sulfur dioxide; there are also methane and bisphenylpropane in products of decomposition at temperatures below 500 °C.

So, the polysulfone is one of the most stable to thermo-oxidation thermoplasts.

The articles from polysulfone possess self-attenuating properties, caused by the nature of polymer but not of the additives. The values of oxygen index for polysulfone lie in the range of 34–38%. Apparently, aromatic polysulfones damp down owing to the formation of carbonized layer, becoming porous protective cover, on their surfaces. The probable is the explanation according to which the inert gas is released from the polymer [8].

Aromatic polysulfone is chemically stable. It is resistant to the effect of mineral acids, alkalis and salt solutions. It is even more stable to carbohydrate oils at higher temperatures and small loadings [149].

The polysulfone on the basis of bisphenylpropane also possesses high dielectric characteristics: volume resistivity -10^{17}; electric inductivity at a frequency 10^6 Hz is 3.1; dielectric loss tangent at a frequency 10^6 Hz is 6×10^{-4}.

Aromatic polysulfones can be reprocesses by die-casting, by means of extrusion, pressing, blow method [150, 151]. The aromatic polysulfones should be dried out for approximately 1.8×10^4 sec at 120 °C until the moisture is more than 0,05%. The quality of articles based on polysulfone worsens at higher moists though the polymer itself does not change its properties. The polysulfones are reprocessible at temperatures 315–370 °C.

The high thermal stability of aromatic polysulfones allows one to conduct multitime reprocessing without the destruction of polymer and loss of properties. The pressed procurements can be mechanically treated on common machine tools.

Besides the aromatic polysulfone on the basis of bisphenylpropane, the other aromatic polysulfones are produced industrially. In particular, the industrial manufacture of polysulfones has been carried out by company "Plastik 3 M" (USA) since 1967 (the polymers are produced by reaction of electrophylic replacements at presence of catalyst). The polysulfone labeled "Astrel-360" contains links of following constitution:

The larger amount of first-type links provides for the high glassing temperature (285 °C) of the polymer. The presence of second-type links allows one to reprocess this thermoplast with help of pressing, extrusion and die casting on special equipment [152].

The company "ICI" (Great Britain) produced polysulfones identical to "Astrel-360" in its chemical structure. However, the polymer "720 P" of that firm contains greater number of second-type links. Due to this, the glassing temperature of the polymer is 250 °C and its reprocessing can be done on standard equipment [153].

Pure aromatic structure of given polymers brings them high thermo-oxidation stability and deformation resistance. Mechanical properties of polysulfone allow one to classify them as technical thermoplasts having high durability, rigidity and impact stability. The polymer "Astrel 360" is in the same row with carbon steel, polycarbonate and nylon on its durability.

The polysulfone has following main characteristics: water absorption – 1.8% [within 5.64 10^4 sec); tensile strength – 90 MPa (20°C) and 30 MPa] (260°C); bending strength – 120 MPa (20°C) and 63 MPa (260°C); modulus in tension – 2.8 GPa (20°C); molding shrinkage – 0.8%; electric inductivity at a frequency 60 Hz – 3.94; dielectric loss tangent at 60 Hz – 0.003 [154–157].

The polysulphones have good resistance to the effect of acids, alkalis, engine oils, oil products and aliphatic hydrocarbons.

The polysulfone "Astrel-600" can be reprocesses at harder conditions. Depending on the size and shape of the article, the temperature of pressed material should lie within the ranger of 315–410°C while pressure should be about 350 MPa [158]. The given polysulfone can be reprocessed by any present method. Articles from it can be mechanically treated and welded. Thanks to its properties, polysulfone finds broad use in electronic, electrotechnique and avia-industry [159, 160].

The papers [161–163] reported in 1968 on the production of heat-resistant polyarylenesulfonoxides "Arilon" of following constitution (company "Uniroyal," USA):

"Arilon" has high rigidity, impact-resistance and chemical stability. The gain weight was 0.9% after 7 days of tests in 20% HCl and 0.5% in 10% NaOH.

This polymer is suitable for long-term use at temperatures 0–130 °C and can be easily reprocessed into articles by die-casting. It also can be extruded and exposed to vacuum molding with deep drawing.

Given polysulfone finds application in different fields of technique as constructional material.

Since 1972, the company "ICI" (Great Britain) has been manufacturing the polyarylenesulfonoxide named "Victrex" which forms at homopolycondensation of halogenphenoles [164–166] of coming structure formula:

Several types of polysulfone "Victrex" can be distinguished: 100P–powder for solutions and glues; 200P–casting polymer; 300P–with increased molecular weight for extrusion and casting of articles operating under the load at increased temperature in aggressive media; and others.

The polysulfone "Victrex" represents amorphous thermoplastic constructional polymer differing by high heat-resistance, dimensional staunchness, low combustibility, chemical and radiation stability. It can easily be reprocessed on standard equipment at 340–380°C and temperature of press-form 100–150 °C. It is dried at 150 °C for 1, 08 * 10^4 sec before casting [167, 168].

This polysulfone is out of wide distribution yet. However, it will, presumably, supplant the part of nonferrous metal in automobile industry, in particular, when producing carburetors, oil-filters and others. The polysulfone successively competes with aluminum alloys in avia industry: the polymer is lighter and yields neither in solidity, neither in other characteristics.

The company "BASF" (Germany) has launched the production of polyethersulfone labeled "Ultrason E" [169] representing amorphous thermoplastic product of polycondensation; it can be characterized by improved chemical stability and fire-resistance. The pressed articles made of it differ in solidity and rigidity at temperature 200 °C. It is assumed to be expedient to use this material when producing articles intended for exposure to increased loadings when the sizes of the article must not alter at temperatures from −100 °C till 150 °C. These items are, for instance in electrotechnique, coils formers, printing and integrated circuits, midspan joints and films for condensers.

"Ultrason E" is used for producing bodies of pilot valves and shaped pieces for hair dryers.

The areas of application of polysulfones are extremely vary and include electrotechnique, car and aircraft engineering, production of industrial, medical and office equipment, goods of household purpose and packing.

The consumption of polysulfones in Western Europe has reached 50% in electronics/electrotechnique, 23% in transport, 12% in medicine, 7% in space/aviation and 4% in other areas since 1980 till 2006 [170].

The polysulfones are used for manufacturing of printed-circuit substrates, moving parts of relays, coils, clamps, switches, pipes socles, potentiometers details, bodies of tools, alkaline storage and solar batteries, cable and capacitor insulation, sets of television and stereo-apparatuses, radomes. The details under bonnet, the head lights mirrors, the flasks of hydraulic lifting mechanisms of cars are produced from polysulfones.

They are also met in internal facing of planes cockpits, protective helmets of pilots and cosmonauts, details of measuring instruments.

The polysulfones are biologically inert and resistant to steam sterilization and γ-radiation what avails people to use them in medicine when implanting artificial lens instead of removed due to a surgical intervention and when producing medical tools and devices (inhalers bodies, ophthalmoscopes, etc.).

The various methods of synthesizing of polysulfones have been devised by Russian and abroad scientists within the last 10–15 years.

The bisphenylolpropane (Diane), 4,4'-dioxybisphenylsulfone, 4,4'-dioxybisphenyl, phenolphthalein, hydroquinone, 4,4'dioxyphenylsulfonylbisphenyl are used as initial monomers for the synthesis of aromatic polysulfones. The polycondensation is carried out at temperatures 160–320 °C with dimethylsulfoxide, dimethylacetamide, N-methylpirrolydone, dimethylsulfone and bisphenylsulfone being used as solvents [171–174].

The Japanese researchers report [175–179] on the syntheses of aromatic polysulfones via the method of polycondensation in the environment of polar dissolvent (dimethylformamide, dimethylacetamide, and dimethylsulfoxide) at 60–400 °C in the presence of alkali metal carbonates within 10 min to 100 h. The synthesized thermoplastic polysulfones possess good melt fluidity [175].

The durable thermo-stable polysulfones of high melt fluidity can be obtained by means of polycondensation of the mix of phenols of 1,3-bis(4-hydroxy-1-isopropylidenephenyl and bisphenol A with 4,4'-dichlordimethylsulfone in the presence of anhydrous potassium carbonate in the environment of dimethylformamide at temperature 166 °C. The solution of the polymer is condensed in MeOH, washed by water and dried out at 150 °C in vacuum. The polysulfone has η_{limit} = 0.5 deciliter/gram (1% solution in dimethylformamide at 25°C) [176].

The aromatic polysulfones with the degree of crystallinity of 36% can be synthesized [179] with help of reaction of 2,2-bis(4-hydroxy-4-tret-butylphenyl) propane with 4,4'-dichlorbisphenylsulfone in the presence of potassium carbonate in the environment of polar solvent 1,3-dimethyl-2-imidasolidinone in current nitrogen at temperatures 130–200 °C. After separation and purification polymer has η_{limit} = 0.5 deciliter/gram.

The possibility to use the synthesized aromatic polysulfone as antipyrenes of textile materials was shown in [180].

The production of polysulfones was shown to be possible in [181] by means of interaction of sulfuric acid, sulfur trioxide or their mixes with aromatic compounds (naphthalene, methylnaphthalene, methoxynaphthalene, dibenzyl ester, bisphenylcarbonate, bisphenyl, stilbene) if one used the anhydride of carbonic acid at 30–200 °C as the process activator. Obtained polymers could be reprocesses by means of pressing.

The synthesis of polyarylenesulfones containing links of 1,3,5-triphenylbenzene can be performed [182] by oxidation of polyarylene thioesters. Obtained polyarylenesulfones have T_{glass} = 265–329 °C and temperature of 5% loss of weight of 478–535 °C. They are well soluble in organic solvents and possess fluorescent properties.

To produce the polysulfone with alyphatic main chain, the interaction of SO_2-group of vinyl-aromatic compound (for example of sterol) with nonsaturated compound (for instance of acrylonitrile) or cyclic olefin (such as 1, 5-cyclooctadiene) has been provided [183] in the presence of initiator of the radical polymerization in bulk or melt, the reaction being carried out at temperatures from 80 to 150 °C.

The method of acceptor-catalyst polyetherification can be used to produce [184, 185] thermo-reactive polyarylenesulfones on the basis of nonsaturated 4,4′-dioxy-3,3′-diallylbisphenyl-2,2′-propane and various chlor- and sulfo- containing monomers and oligomers. The set of physic-mechanical properties of polymers obtained allows one to propose them as constructional polymers, sealing coatings, film materials capable of operating under the influence of aggressive environments and high temperatures.

Relatively few papers are devoted to the synthesis and study of polycondensational block-copolyesters. At the same time this area represents, undoubtedly, scientific and practical interest. The number of papers [186–190] deals with the problem of synthesis and study of physic-chemical properties of polysulfones of block constitution.

For example, it is reported in [186] on the synthesis of block-copolyesters and their properties in dependence on the composition and structure of oligoesters. The oligoesters used were oligoformals on the basis of Diane with the degree of condensation 10 and oligosulfones on the basis of phenolphthalein with the degree of condensation 10. The synthesis has been performed in conditions of acceptor-catalyst polycondensation.

The polysulfone possesses properties of constructional material: high solidity, high thermo-oxidation stability.

However, the deficiency of polysulfonic material is the high viscosity of its melt what results in huge energy expenses at processing. One can decrease the viscosity if, for example, "sews" together polysulfonic blocks with liquid-crystal nematic structures which usually have lower values of viscosity. To produce the block-copolymer, the flexible block of polysulfone and rigid polyester block (with liquid-crystal properties) were used in Ref. [187]. The polysulfonic block was used as already ready oligomer of known molecular weight with edge functional groups.

Polyester block represented the product of polycondensation of phenylhydroquinone and terephthaloylchloride.

The synthesis of block-copolymer went in two stages. At first, the polyester block of required molecular weight was produced while on the second stage that block (without separation) reacted to the polysulfonic block. The synthesis was carried out by the method of high-temperature polycondensation in the solution at 250 °C in environment of α-chlornaphthalene. The duration of the first stage was 1.5 hour, and second stage was 1 hour. Obtained polymers exhibited liquid-crystal properties.

The block-copolymers of polysulfone and polyesters can be also produced [188] by the reaction of aromatic polysulfones in environment of dipolar aprotone dissolvents (dimethylsulfoxide, N-methylpirrolydone, N-methylcaprolactam, N, N{}-dimethylacetamides or their mixes) with alyphatic polyesters containing not less than two edge OH-groups in the presence of basic catalyst – carbonates of alkali metals: Li, Na, K.

Polysulfone is thermo-mechanical and chemically durable thermoplast. But in solution, which is catalyzed by alkali, it becomes sensitive to nucleophylic replacements. In polar aprotone dissolvents at temperatures above 150 °C in the presence of spirit solution of K_2CO_3 it decomposes into the bisphenol A and diarylsulfonic simple esters. The analogous hydrolysis in watery solution of K_2CO_3 goes until phenol products of decomposition. This reaction is of preparative interest for the synthesis of segmented block-copolyester simple polyester – polysulfone. The transetherification of polysulfone, being catalyzed by alkali, results, with exclusion of bisphenol A and introduction segments of simple ester, in formation of segmented block-copolymer [189].

The Ref. [190–199] are devoted to the problem of synthesis of statistical copolymers of polysulfones, production of mixes on the basis of polysulfones and study of their properties as well to the mechanism of copolymerization.

The graft copolymer products, poly (met)acrylates branched to polyestersulfones, can be produced next way [200]. Firstly, the polyestersulfone is being chlormethylenized by monochlordimethyl ester. The product is used as macrostarter for the graft radical polymerization of methylmethacrylate (I), methylacrylate (II) and butylacrylate (III) in dimethylformamide according to the mechanism of transferring of atoms under the influence of the catalytical system $FeCl_2$/isophthalic acid. The branched copolymer with I has only one glassing temperature while copolymer with II and III has three.

Lately, several papers on the synthesis of liquid-crystal polysulfones and on the production of mixes and melts of polysulfones with liquid-crystal polymers have appeared [201–205].

For example, the polysulfone "Udel" can be modified [201] by means of introduction of chlormethyl groups. Then the reaction of transquarternization of chlormethylized polysulfone and obtained nitromethylene dimesogens is conducted. The dimesogens contain one phenol OH-group and form nematic mesophase in liquid state. Obtained such a way liquid-crystal polysulfones, having lateral rigid dimesogenic links possess the structure of enantiotropic nematic mesophase fragments.

The Ref. [202] reports on the synthesis of liquid-crystal polysulfone with mesogenic link cholesterylpentoatesulfone.

The effect of compatibility, morphology, rheology, mechanical properties of mixes of polysulfones and liquid-crystal polymers are studied in Refs. [203–205]. There are several contributions on the methods of synthesizing of copolymers of polysulfones and polyesterketones and on the production of mixes [206–210]. The method for the synthesis of aromatic copolyestersulfoneketones proposed in Ref. [206] allows one to decrease the number of components used, to lower the demands to the concentration of moist in them and to increase the safety of the process. The method is in the interaction without aseotropoformer in environment of dimethylsulfone of bisphenols, dihaloydarylenesulfones and (or) dihaloydaryleneketones and alkali agents in the shape of crystallohydrated of alkali metal carbonated and bicarbonates. All components used are applicable without preliminary drying.

The novel polyarylenestersulfoneketone containing from the cyclohexane and phthalasynone fragments is produced via the reaction of nucleophylic replacement of 1-methyl-4, 5-bis (4-chlorbenzoyl) cyclohexane, 4, 4'-dichlorbisphenylsulfone and 4-(3, 5-dimethyl-4-hydroxyphenyl)-2,3 phthalasine-1-one. The polymer is described by means of IR-spectroscopy with

Fourier-transformation, ^{1}H nuclear magnetic resonance, differential scanning calorimetry and diffraction of X-rays. It is shown that the polymer is amorphous and has high glassing temperature (200 °C) it is soluble in several dissolvents at room temperature. The by-products are the "sewn" and graft copolymers [207].

The block copolymers of low-molecular polyesterketoneketone and 4,4'{}-bisphenoxybisphenylsulfone (I) can be produced [208] by the reaction of polycondensation. The glassing temperatures increase and melting points lower of copolymers if the concentration of (I) increases. Block copolymers, containing 32.63–40.7% of (I) have glassing temperature and melting point 185–193 and 322–346°C, respectively; the durability and modulus in tension are 86.6–84.2 MPa and 3.1–3.4 GPa, respectively; tensile elongation reaches 18.5–20.3%. Block-copolymers possess good thermal properties and reprocessibility in melt.

The analysis of literature data and patent investigation has revealed that the production of such copolymers as polyestersulfoneketones, liquid-crystal polyestersulfones as well as the preparation of the mixes and melts of polysulfones with liquid-crystal polyesters of certain new properties are of great importance.

With the account for the upper-mentioned, we have synthesized polyestersulfones on the basis of oligosulfones and terephthaloyl-bis (*n*-oxybenzoic) acid [174, 186, 434] and polyestersulfoneketones on the basis of aromatic oligosulfoneketones, mixes of oligosulfones with oligoketones of various constitution and degree of polycondensation and different acidic compounds [211, 212, 435].

7.6 AROMATIC POLYSULFONES

The stormy production development of polymers, containing aromatic cycles like, for example, the polyarylenesterketones has happened within the past decades. The polyarylenesterketones represent the family of polymers in which phenylene rings are connected by the oxygen bridges (simple ester) and carbonic groups (ketones). The polyarylenesterketones include polyesterketone, polyesteresterketone and others distinguishing in the sequence of elements and ratio E/K (of ester groups to ketone ones). This ratio influences on the glassing temperature and melting point: the higher content of ketones increases both temperatures and worsens reprocessibility [213].

The elementary units of polyesteresterketones contain two simple ester and one ketone groups, while those of polyesterketone only one ester and one ketone [214].

ПЭК

ПЭЭК

Polyesteresterketone is partially crystalline polymer the thermo-stability of which depends on glassing temperature (amorphosity) and melting point (crystallinity) and increases with immobilization of macromolecules. The strong valence bonds define the high thermo-stability and longevity of mechanical and electrical properties at elevated temperature.

The polyesterketone labeled "Victrex^R" was firstly synthesized in Great Britain by company "Imperial Chemical Industries" in 1977. The industrial manufacture of polyesteresterketones started in 1980 in Western Europe and USA and in 1982 in Japan [215].

The polyesteresterketone became the subject of extensive study from the moment of appearance in industry. The polyesteresterketone possesses the highest melting point among the other high-temperature thermoplasts (335 °C) and can be distinguished by its highly durable and flexible chemical structure. The latter consists of phenylene rings, consecutively joined by para-links to ester, ester and carboxylic groups. There is a lot of information available now on the structure and properties of polyesteresterketones [216, 217].

The polyesteresterketone is specially designed material meeting the stringent requirements from the point of view of heat resistance, inflammability, products combustions and chemical resistance [218, 219]. "Victrex^R" owns the unique combination of properties: thermal characteristics and combustion parameters quite unusual for thermoplastic materials, high stability to effect of different dissolvents and other fluids [220, 221].

The polyesteresterketone can be of two types: simple (unarmored) and reinforced (armored) by glass. Usually both types are opaque though they can become transparent after treatment at certain conditions. This happens due to the reversible change of material's crystallinity, which can be recovered by tempering. The limited number of tinges of polyesteresterketones has been produced for those areas of industry where color articles are used.

The structure crystallinity endows polyesteresterketones by such advantages as

- stability to organic solvents,
- stability to dynamical fatigue,
- improved thermal stability when armoring with glass,
- ability to form plasticity at short-term thermal aging,
- orientation results in high strength fibers.

"Victrex^R" loses its properties as elasticity and solidity moduli with increasing temperature, but the range of working temperatures of polyesteresterketones is wider in short-term process (like purification) than that of other thermoplastic materials. It can be exploited at 300 °C or higher. The presumable stint at 250 °C is more than 50,000 hours for the given polyesteresterketone. If one compares the mechanical properties of polyesteresterketone, polyestersulfone, nylon and polypropylene then he finds that the first is the most resistant to wear and to dynamic fatigue. The change of mechanical characteristics was studied in dependence of sorption of CH_2CCl_2.

Polyesteresterketones, similar to any other thermoplastics, is isolation material. It is hard-burnt and forms few smoke and toxic odds in combustion, the demand in such materials arises with all big hardening requirements to accident prevention.

If one compares the smoke-formation in combustion of 2–3 mm of samples from ABC-plastics, polyvinylchloride, polystyrene, polycarbonate, polytetrafluorethylene, phenolformaldehyde resin, polyestersulfone, polyesteresterketone then it occurs that the least smoke is released by polyesteresterketone, while the greatest amount of smoke is produced by ABC-plastic.

"Victrex^R" exhibits good resistance to water reagents and pH-factor of different materials starting with 60% sulfuric acid and 50% potassium hydroxide. The polyesteresterketone dissolves only in proton substances (such as concentrated sulfuric acid) or at the temperature close to its melting point. Only α-chlornaphthalene (boiling point 260 °C) and benzophenone influence on "Victrex^R" among organic dissolvents.

The data on the solubility have revealed that two classes of polyester-esterketones coexist: "amorphous" and crystalline [222].

The division of these polymers into two mentioned classes is justified only by that the last class, independent of condensation method, crystallizes so fast in conditions of synthesis that the filtering of combustible solution is not possible. It may be concluded from results obtained that "amorphous" class of polyesteresterketones is characterized by bisphenols, which have hybridized sp^3-atom between phenyl groups.

From the point of view of short-term thermal stability the polyesteresterketones do not yield most steady materials polyestersulfones destruction of which is 1% at 430 °C. Yet and still their long-term stability to UV – light, oxygen and heat must be low due to ketone-group [222].

The influence of environment on polyesteresterketones is not understood in detail, but it has proven that polyesteresterketone fully keeps all its properties within 1 year. Polyesteresterketones exhibit very good stability to X-ray, β- and γ- radiations. Wire samples densely covered with polyesteresterketones bear the radiation 110 Mrad without essential destruction.

The destroying tensile stress of polyesteresterketone is almost nil at exposure to air during 100 hours at 270 °C. At the same time the flex modulus at glassing temperature of 113 °C falls off precipitously, however remains sufficiently high compared to that of other thermoplasts.

When placed into hot water (80 °C) for 800 hours the tensile stress and the permanent strain after rupture of polyesteresterketones decreases negligibly. The polyesteresterketone overcomes all the other thermoplasts on stability to steam action. The articles from polyesteresterketone can stand short exposure to steam at 300 °C.

On fire-resistance this polymer is related to hard-to-burn materials.

The chemical stability of polyesteresterketone "Victrex^R" is about the same as of polytetrafluorethylene while its long-term strength and impact toughness are essentially higher than those or nylon A-10 [223].

The manufacture of the polyesteresterketones in Japan is organized by companies "Mitsui Toatsu Chem" under the labels "Talpa-2000", "ICI Japan," "Sumitoma Kogaku Koge." The Japanese polyesteresterketones have glassing temperature 143 °C and melting point 334 °C [224, 225].

The consumption of polyesteresterketones in Japan in 1984 was 20tons, 1 kilogram cost 17000Ian. The total consumption of polyestersulfones and polyesteresterketones in Japan in 1990 was 450–500 tons per year [226].

Today, 35% of polyesteresterketones produced in Japan (of general formula [$-OC_6H_4-O-C_6H_4-CO-C_6H_4-]_n$) are used in electronics and electrotechnique, 25% in aviation and aerospace technologies, 10% in car manufacture, 15% in chemical industry as well as in the fields of everyday life, for example at producing the buckets for hot water, operating under pressure and temperature up to 300 °C. The Japanese industrial labels of polyesteresterketones have good physical-mechanical characteristics: the high impact toughness; the heat resistance (152 °C, and 286 °C with introduction of 20% of glass fiber); the chemical resistance (bear the influence of acids and alkalis, different chemicals and medicines), the tolerance for radiation action; the elasticity modulus 250–300 kg/mm^2; the rigidity; the lengthening of 100%; the negligible quantity of smoke produced. The polyesteresterketone can be reprocessed by die casting at 300–380 °C (1000–1400 kg/cm^2), extrusion, formation and others methods [224]. High physical-mechanical properties remain the dame for the long time and decrease on 50% only after 10 years.

The polymer is expensive. One of the ways for reducing the product price is compounding. The company "Kogaku Sumitoma" created compounds "Sumiploy K" on the basis of polyesteresterketones by their original technology. The series "Sumiploy K" includes the polymers with high strength and of improved wear resistance. The series "Sumiploy SK" is based on the polyesteresterketone alloyed with other polymers [224]. The series includes polymers, the articles of which can be easily taken out from the form, and products are of improved wear resistance, increased high strength, good antistatic properties, can be easily metalized.

The company "Hoechst" (Germany) produces unreinforced polyesteresterketones "Hoechst X915," armored with 30 wt% of glass fiber (X925) and carbon fiber (X935), which are characterized by the good physical-mechanical properties (Table 7.4). The unreinforced polymer has the density almost constant till glassing temperature (about 160 °C). The reinforcing with fibers permits further to increase the heat resistance of polyesteresterketone. At the moment the polyesteresterketones labeled "Hostatec" are being produced with 10, 20 and 30 wt.% of glass and carbon fibers. Several labels of polyesteresterketones are under development which involve mineral fillers, are not reinforced and contain 30 wt.% of glass and carbon fibers.

The constructional thermoplast "Hostatec" dominates polyoxymethylene, polyamide and complex polyesters on many parameters.

The polyesteresterketones can be easily processed by pressing, die cast and extrusion. They can be repeatedly crushed to powder for secondary utilization. They are mainly used as constructional materials but also can be used as electroinsulation covers operating at temperatures of 200 °C and higher for a long time [225].

TABLE 7.4 Physic-Mechanical Properties of Polyesteresterketones of Company "Hoechst"

Property	Unreinforced	Containing 30 weight %	
		Glass fiber	Carbon fiber
Density, g/cm^3	1.3	1.55	1.45
Linear shrinkage, %	1.5	0.5	0.1
Tensile strength, N/mm^2	86	168	218
Breaking elongation, %	3.6	2.2	2.0
Modulus in tension, kN/mm^2	4	13.5	22.5
Impact toughness notched, J/m	51	71	60
Heat resistance, °C	160	Above 320	320

The polyesteresterketone found application in household goods (in this case its high heat resistance and impact toughness are being used), in lorries (joint washers, bearings, probes bodies, coils and other details, contacting fuel, lubricant and cooling fluid).

The big attention to polyesteresterketones is paid in aircraft and space industries. The requirements to fire-resistance of plastics used in crafts have become stricter within the last years. Unreinforced polyesteresterketones satisfy these demands having the fire-resistance category U–O on UL 94 at thickness 0.8 mm. In addition, this polymer releases few smoke and toxic substances in combustion (is used in a subway). The polyesteresterketone is used for coating of wires and cables, used in details of aerospace facility (the low inflammability, the excellent permeability and the wear stability), in military facility, ship building, on nuclear power plants (resists the radiation of about 1000 Mrad and temperature of water steam 185 °C), in oil wells (pillar stand to the action of water under pressure, at a temperature of 288 °C), in electrical engineering and electronics.

Polyesteresterketone offer properties of thermoreactive resin, can be easily pressed, undergoes overtone, and resists the influence of alkalis.

Since January 1990, the Federal aviation authority have adopted the developed at Ohio State University method, in which the heat radiation (HR) and the rate of heat release (RHR) are determined. The standard regulates the HR level at 65 kJ. Many aircraft materials do not face this demand: for example, the ternary copolymer of acrylonitrile, butadiene and styrene resin, the polycarbonate, phenol and epoxide resins. It is foreseen to substitute these materials by such polymers, which meet these requirements. The measurements in combustion chamber of Ohio State University indicated that polyesteresterketone fulfills this standard.

The heightened activity in creating and evaluation of composite properties on the basis of polyesteresterketones occurs recent years [228]. The thermoplastic composites have the number of advantages regarding the plasticity, maintainability and ability for secondary utilization compared to epoxy composites. These polymers are intended to be applied in alleviated support elements. In the area of cable insulation, polyesteresterketones can be reasonably used when thermo-stability combined with fire-resistance without using of halogen antipyrenes where is desired.

The "Hostatec" has low water absorption. Dielectric properties of films from polyesteresterketone "Hostatec" are high. This amorphous polymer has the electric inductivity 3, 6, loss factor 10^{-3} and the specific volume resistance 10^{17} Ohm cm, these values remain still up to 60 °C.

Growth of demand for polyesteresterketones is very intense. In connection with growing demands on heat resistance and stability to various external factors the polyesteresterketones find broader distribution. The cost of one kilogram of such polymer is 5–20 times larger than the cost of usual constructional polymers (polycarbonates, polyamides, and polyformaldehydes). But, despite the high prices the polyesteresterketones and compositions on their basis, owing to the high level of consumer characteristics, find more and more applications in all industries. The growth of production volumes is observed every year.

It is known, that the paces of annual growth of polyesteresterketones consumption were about 25% before 1995, and its global consumption in 1995 was 4000 tons. The polyesteresterketone do not cause ecological problems and is amenable to secondary processing.

In connection with big perceptiveness of polyarylesterketones, the examining of the most popular methods for their production was of interest.

The literature data analysis shows that the synthesis of aromatic polyesterketones can be done by the acylation on Friedel-Crafts reaction or by reaction of nucleophylic substitution of activated dihalogen-containing aromatic compounds and bisphenolates of alkali metals [224, 229].

In the majority cases, the polyesterketones and polyesteresterketones are produced by means of polycondensation interaction of bisphenols with 4,4′-dihalogen-substituted derivatives of benzophenone [230–263], generally it is 4,4′-difluoro-or dichlorbisphenylketone. The introduction of replacers into the benzene ring of initial monomer raises the solubility of polyesterketones and polyesteresterketones. So, polyesterketone on the basis of 3,3′,5,5′-tetramethylene and 4,4′-difluorobenzophenone is dissolved at 25 °C [211–214] in dimethylsulfoxide, the reduced viscosity of melt with concentration 0.5 gram/deciliter is 0.79 deciliter/gram.

The polyesterketone based on 4, 4′-dioxybenzophenone and dichlormethylenized benzene derivatives is soluble in chloroform and dichloroethane [264]. The logarithmic solution viscosity of polyesterketone obtained on the basis of 4,4′-dioxybisphenylsulfone and 4,4′-dichlorbenzophenone in tetrachloroethane of concentration 0.5 gram/deciliter is 0.486; the film materials from this polyester (of thickness 1 mm) are characterized by the high light transmission (86%) and keep perfect solubility and initial viscosity after exposure to 320 °C during 2hours.

The high-boiling polar organic solvents dimethylsulfoxide, sulfolane, dimethylsulfone, dimethylformamide, dimethylacetamide are generally used for synthesizing the polyesterketones and polyesteresterketones by means of polycondensation; in this case the reaction catalysts are the anhydrous hydroxides, carbonates, fluorides and hydrides of alkali metals. The polymers synthesis is recommended to be carried out in inert gas atmosphere at temperatures 50–450 °C. If catalysts used are the salts of carbonic or hydrofluoric acids then oligomers appear. Chain length regulators when producing polyesterketones based on difluoro-or dichlorbenzophenone and bisphenolates of alkali metals and dihydroxynaphthalenes can be the monatomic phenols [251–253].

The synthesis of polyesterketones and polyesteresterketones according to the Friedel-Crafts reaction is lead in mild conditions [265–283]. So, solidifying thermo-stable aromatic polesterketonesulfones, applied as binding agents when laminating, can be produced [278] in the presence of aluminum chloride by the interaction of 1,4-di(4-benzoylchloride)butadiene-1,3-dichloranhydrides of iso- and terephthalic acids, bisphenyl oxide and 4,4'-bisphenoxybisphenylsulfone.

The aromatic polyesterketones and their thioanalogs are synthesized [266–281, 284] with help of polycondensation of substituted and not-substituted aromatic esters and thioesters with choric anhydrides of dicarboxylic acids in environment of aprotone dissolvents at temperatures from −10 till 100 °C in the presence of Lewis acids and bases.

The aromatic polyesterketones and polyesterketonesulfonamides based on 4,4'-dichloranhydride of bisphenyloxidebicarbonic acid and 4-phenoxybenzoylchloride can be produced by means of Friedel-Crafts polycondensation in the presence of AlCl$_3$ [270]. The reduced viscosity of the solution in sulfuric acid of concentration 0.5 gram/deciliter is 0.07–1.98 deciliter/gram. The Friedel-Crafts reaction can also be applied to synthesize the copolyesterketones from bisphenyl ester and aromatic dicarboxylic acids or their halogenanhydrides [274]. The molecular mass of polymers, assessed on the parameter of melt fluidity, peaks when bisphenyl ester is used in abundant amount (2–8%).

The polyarylesterketones can be produced by means of interaction between bisphenylsulfide, dibenzofurane and bisphenyloxide with monomers of electrophylic nature (phosgene, terephthaloylchloride) or using homopolycondensation of 4-phenoxybenzoylchloride and 4-phenoxy-4-chlorcarbonyl-bisphenyl in the presence of dichloroethane at 25 °C [282–285]. Aromatic polyesterketones form after the polycondensation of 4-phenoxybenzoylchloride with chloranhydrides of tere- and isophthalic acids, 4,4' dicarboxybisphenyloxide in the environment of nitrobenzene, methylchloride and dichloroethane at temperatures from 70 till 40 °C during 16–26 h according Friedel-Crafts reaction.

The synthesis of polyesterketones based on aromatic ether acids is possible in the environment of trifluoromethanesulfonic acid [249, 287]. The data of ^{13}C nuclear magnetic resonance has revealed [249] that such polyesterketone comprise only the n-substituted benzene rings. When using the N-cyclohexyl-2-pyrrolidone as a solvent when synthesizing polyphenylenesterketones and polyphenylenethioesterketones the speed of polycondensation and the molecular mass of polymers [288] increase.

So, the polymer is produced, the reduced viscosity of 0.5% solution of which in sulfuric acid is 1.0 deciliter/gram, after the interaction of 4,4'-difluorobenzophenone and hydroquinone at 290 °C in the presence of potassium carbonate during 1 hour. Almost the same results are obtained when synthesizing polyphenylenethioesterketones. However, the application of the mix bisphenylsulfone/sulfolane as a solvent during the interaction of 4,4' difluorobenzophenone with sodium sulfide within 2–13 h provides for the production of polymer with reduced viscosity 0.23–0.25 deciliter/gram. The high-molecular polyarylenesulfideketones, suitable for the preparation of films, fibers and composite materials are formed during the interaction of 4,4' dihalogenbenzophenone with sodium hydrosulfide in solution (N-methylpirrolydone) at 175–350 °C within 1–72 h [289].

For the purpose of improving physical-mechanical properties and increasing the reprocessibility of polyesterketones and polyesteresterketones their sulfonation by liquid oxide of hexavalent sulfur in environment of dichloroethane has been carried out [290]. In this case the polymer destruction does not happen, which is observed at sulfonation by concentrated sulfuric or chlorosulfonic acid. The abiding flexible film materials can be produced from sulfonated materials using the method of casting. The aromatic polyesterketones could also be produced by oxidative dihydropolycondensation (according to the Scolla reaction) of 4,4'-bis-(1-naphthoxy)-benzophenone at 20 °C in the presence of trivalent iron chloride in environment of nitrobenzene [291]. The mix of pentavalent phosphorus oxide and methylphosphonic acid in ratio 1:10 can be used as a solvent and dehydrate agent [292, 293] when synthesizing the aromatic polyesterketones on the basis of bisphenylester of hydroquinone, 4,4'-bisphenyloxybicarbonic acid, 1,4-bis(m-carboxyphenoxy) benzene, and also for homopolycondensation of 3- or 4-phenoxybenzoic acid at 80–140 °C.

The sulfur-containing analogs of polyesterketones, that is, polythioesterketones and copolythioesterketones can be synthesized [294–303] by polycondensation of dihalogenbenzophenols with hydrothiophenol or other bifunctional sulfur-involving compounds, and also of their mixes with different bisphenols in environment of polar organic solvents. As in case of polyesterketones, the synthesis of their thioanalogs is recommended to be carried to out in inert medium at temperatures below 400 °C in the presence of catalyst (hydroxides, carbonates and hydrocarbonates of alkali metals).

The aromatic polyesterketones can also be produced by polycondensation or homopolycondensation of compounds like halogen-containing arylketonephenols and arylenedihalogenides of different functionality at elevated temperature and when the salts of alkali and alkali-earth metals in environments of high-boiling polar organic solvents are used as catalysts

[222, 304–312]. The 4-halogen-3-phenyl-4-hydroxybenzophenone, 4-(n-haloidbenzoyl)-2, 6-dinethylphenol and others belong to monomers with mixed functional groups.

The polycondensation process when synthesizing polyesterketones and polyesteresterketones can be performed in the melt [223, 313–317] too. So, it is possible to produce the aromatic polyesterketones by means of interaction in melt of 4,4′ difluorobenzophenone with trimethylsiloxane esters of bisphenols with different bridged groups in the presence of catalyst (cesium fluoride – 0.1% from total weight of both monomers) at 220–270 °C [318]. The monomers do not enter the reaction without catalyst at temperatures below 350 °C. The reduced viscosity of 2% solution of polymer in tetrachloroethane at 30 °C is 0.13–1.13 deciliter/gram, the molecular weight 3200–60,000, glassing temperature is 151–186 °C, melting point is 240–420 °C. According to the data of thermogravimetric analysis in the air the mass loss of polymer is less than 10%, when temperature elevates from 422 °C to 544 °C with the rate 8 degrees per minute.

To increase the basic physical-mechanical characteristics and reprocessibility (the solubility, in particular), the polyesterketones and polyesteresterketones are synthesized [319–329] through the stage of formation of oligomers with end functional groups accompanied with the consequent production of block-copolyesterketones or by means of one-go-copolycondensation of initial monomers with production of copolyesterketones.

The polyesteresterketones, their copolymers and mixes are used for casting of thermally loaded parts of moving transport, instrument, machines, and planes. They are used in articles of space equipment: for cable insulation, facings (pouring) elements. For instance, the illuminators frames of planes and rings for high-frequency cable are made of polyesterketone "Ultrapek." It is widely used in electronics, electrician, for extrusion of tubes and pipes operating in aggressive media and at low temperatures. The polyesteresterketones and polyesterketones are used for multilayer coating as the basis of printed planes. The conservation of mechanical strength in conditions of high humidity and temperature, the stability to radiation forwards their application to aerospace engineering.

The compositions on the basis of polyesteresterketones already compete with those based on thermoreactive resins when making parts of military and civilian aircrafts. Sometimes it is possible to cut the weight on 30% if produce separate parts of plane engines from reinforced polyesteresterketones.

It is proposed to use the polyesteresterketones in the manufacture of fingers of control rod and cams of brake system, motors buttons in car industry. The piston cap of automobile engine, made from polyesterester-ketone "Victrex," went through 1300 h of on-the-road tests.

The advantage of using this material, contrary to steel, lays in wearout decrease, noise reduction, 40% weight reduction of the article. The important area of application of polyesteresterketones can be the production of bearings and backings. The polyesteresterketones are recommended to be used for piece making of drilling equipment (zero and supporting mantles) and timber technique, in different joining's of electric equipment of nuclear reactors, layers and valves coverings, components of sports facility.

The fibers of diameter 0.4 mm (from which the fabric is weaved in shape of tapes and belts used in industrial processes where the temperature-resistant, the high-speed conveyers are needed) are produced from polyesteresterketone melt at temperatures of 350–390 °C. The fabric from polyesteresterketone or polyesterketone keeps 90% of the tensile strength after thermal treatment at 260 °C, does not change its properties after steaming at 126 °C for 72 h under the load, and resists alkalis action with marginal change. The medical instruments, analytical, dialysis devices, endoscopes, surgical and dental tools, containers made of polyarylesterketones can be sterilized by steam and irradiation [330].

The manifold methods of synthesizing polyarylenesterketones have been devised by Russian and foreign scientists within the last 20 years.

For example, the method for production of aromatic polyketones by means of Friedel-Craft polycondensation of bis (arylsilanes) with chlorides of aromatic dicarboxylic acids (isophthaloyl-,terephthaloyl-,4,4{}-oxydibenzoylchloride) at 20 °C in environment of dissolvent (1,2-dichlorethane) in the presence of aluminum chloride is proposed in Ref. [331]. The polyketones have the intrinsic viscosity more than 0.37 deciliter/gram (at 30 °C, in concentrated H_2SO_4), glass transition temperature is 120–231 °C and melting point lies within 246–367 °C. The polyketones start decomposing at a temperature of 400 °C, the temperature of 10% loss of mass is 480–530 °C.

It is possible to produce polyketones by the reaction of aromatic dicarboxylic acid and aromatic compound containing two reactive groups [332]. The reaction is catalyzed by the mix of phosphoric acid and carboxylic acid anhydride having the formula of RC(O)O(O)CR (R stands for not-substituted or substituted alkyl, in which one, several or all hydrogen atoms were replaced by functional groups and each R has the Gamet constant $\sigma_m \geq 0.2$). The pressed articles can be created from synthesized polyketones.

The aromatic polyketones can be synthesized [333] when conducting the Friedel-Craft polymerization with acylation by means of the interaction of 2,2{} bis (arylphenoxy) bisphenyls with chlorides of arylenebicarbonic acids in the presence of $AlCl_3$. The polyketones obtained using the most efficient 2,2{} bis (4-bezoylphenoxy) bisphenyl are well soluble in organic solvents and possess high heat resistance.

Also, the aromatic polyketones are produced [334] by the reaction of electrophylic substitution (in dispersion) of copolymer of aliphatic vinyl compound (1-acosen) with N-vinylpirrolydone, at ratio of their links close to equimolar.

The other method of synthesizing aromatic polyketones includes the interaction of monomer of formula [HC (CN) (NR$_2$)] Ar with 4,4' difluorobisphenylsulfone in dimethylsulfoxide, dimethylformamide or N-methylpirrolydone at 78–250 °C in the presence of the base [335]. The soluble polyaminonitrile of formula [C (CN) (NR$_2$) ArC (CN) (NR$_2$) C6H-4SO$_2$C6H4]$_n$ is thus produced (Ar denotes m-C$_6$H$_4$; NR$_2$ stands for group -NCN$_2$CH$_2$OCH$_2$CH$_2$). When acid hydrolysis of polyaminonitrile is carried out the polysulfoneketone of formula [COArCOC$_6$H$_4$SO$_2$C$_6$H$_4$]$_n$ forms with glassing temperature of 192 °C, melting point 257 °C and when 10% mass loss happens at 478 °C. The acid hydrolysis can be performed in the presence of n-toluenesulfoacid, trifluoroacetic acid, and mineral acids.

The synthesis of aromatic polyketone particles has been carried out [336] by means of precipitation polycondensation and is carried out at very the low concentration of monomer [0, 05 mol/liter]. The polyketones are produced from bisphenoxybenzophenone (0,005 mole) or isophthaloylchloride (0.005 mole) in 100 milliliters of 1,2-dichloroethane. Some of obtained particles have highly organized the needle-shaped structure (the whisker crystals). The use of isophthaloyl instead of terephthaloyl at the same low concentration of monomer results in formation of additionally globular particles, the binders of strip structures gives rise. The average size of needle-shaped particles is 1–5 mm in width and 150–250 mm in length.

The synthesis of aromatic high-molecular polyketones by the low-temperature solid-state polycondensation of 4,4'bisphenoxybenzophenone and isophthaloylchloride in the presence of $AlCl_3$ in 1,2-dichlorethane is possible [337]. Obtained polymers are thermoplasts with glass transition temperature 160 °C and melting point 382 °C.

The polyesterketones are synthesized [338] from dichlorbenzophenone and Na_2CO_3 in the presence of catalyst SiO_2-Cu-salt. The polymer has the negligible number of branched structures and differs from polyester-ketones synthesized on the basis of 4-hydroxy-4'-flourobenzophenone on physical properties (its pressed samples have higher crystallinity and orientation).

There exist reports on syntheses of fully aromatic polyketones without single ester bonds [339–349].

The fully aromatic polyketones without ether bonds were produced [339] on the basis of polyaminonitrile, which was synthesized from anions of bis(aminonitrile) and 4,4'{}-difluorobenzophenone using the sodium hydride in mild conditions. The acid hydrolysis of synthesized polyami-nonitrile avails one to obtain corresponding polyketone with high thermal properties and tolerance for organic solvents.

They are soluble only in strong acids such as concentrated H_2SO_4. The polyketones have glassing temperature 177–198 °C. Their melting points and temperature of the beginning of decomposition are respectively 386–500 °C and 493–514 °C [340].

The aromatic polyketones without ester bonds can also be produced by the polymerization of bis (chlorbenzoyl) dimethoxybisphenyls in the presence of nickel compounds [341]. The polymers have the high molecular mass, the amorphous structure, the glass transition temperature 192 °C and 218 °C and form abiding flexible films.

The known aromatic polyketones (the most of them) dissolve in strong acids, or in trifluoroacetic acid mixed with methylenechloride, or in tri-fluoroacetic acid mixed with chloroform.

It is known that the presence of bulk lateral groups essentially improves the solubility of polyketones, and also improves their thermo-stability. In connection to this, dichloranhydride of 3,3-bis-(4'carboxyphenyl) phtalide (instead of chloranhydride) was used as initial material for condensation with aromatic hydrocarbons [350]. According to the Friedel-Crafts reaction of electrophylic substitution in variant of low-temperature precipitation polycondensation the high-molecular polyarylenepthalidesterketones have been synthesized. Obtained polymers have greater values of intrinsic viscosity 1.15–1.55 deciliter/gram (in tetrachloroethane). The softening temperatures lie within 172–310 °C, the temperature of the beginning of decomposition is 460 °C. Polyarylenepthalidesterketones dissolve in wide range of organic solvents, form colorless, transparent, abiding (σ = 85–120 MPA) and elastic (ε = 80–300%) films when formed from the melt.

The soluble polyketones-polyarylemethylketones can be produced [351] by condensation polymerization of 1, 4-dihalogenarenes and 1, 4-diacetylbenzols in the presence of catalytical palladium complexes, base and phosphoric ligands. The high yield of polymer is seen when using the tetrahydrofurane, o-dichlorbenzole and bisphenyl ester as solvents. The synthesized polymers dissolve in tetrahydrofurane, dichloromethane and hexane has the decomposition point 357 °C (in nitrogen). It has luminescent properties: emits the green light (490–507 nm) after light irradiation with wavelength of 380 nm.

The works [352–355] report on syntheses of carding aromatic polyketones.

The ridge-like polyarylesterketones have been synthesized by means of one-stage polycondensation in solution of bis (4-nitrophenyl) ketone with phenolphthalein, o-cresolphthaleine, 2, 5, 2'5' tctramcthylphenolphthaleine and timolphthaleine [352]. Authors have shown that the free volume within the macromolecule depends on position, type and number of alkyl substitutes.

The homo and copolyarylenesterketones of various chemical constitution (predominantly, carding ones) have been produced by the reaction of nucleophylic substitution of aromatic activated dihaloid compound [353], and also the "model" homopoyarylenesterketones on the basis of bisphenol, able to crystallize, created. The tendency to crystallization is provided by the combination of fragments of carding bisphenols with segments of hydroquinone (especially, of 4, 4'dihydroxydiphenyl) and increases with elongation of difluoro-derivative (the oligomer homologues of benzophenone), and also in the presence of bisphenyl structure in that fragment. Owing to the presence of carding group, the glass transition temperature of copolymer reaches 250 °C, and the melting point 300–350 °C.

The crystallizing carding polyarylenesterketones, in difference from amorphous ones, dissolve in organic solvents very badly, they are well soluble in concentrated sulfuric acid at room temperature and when heating to boiling in m-cresol (precipitate on cooling) [354, 355].

There are contributions [356, 357] devoted to the methods of synthesis of aromatic polyketones on the basis of diarylidenecycloalkanes.

The polyketones are produced [356] by the reaction of 2,7-dibenzyli-denecyclopentanone (I) and dibenzylideneacetone (II) with dichlorides of different acids (isophthalic, 3,3{}-azodibenzoic and others) in dry chloromethane in the presence of $AlCl_3$. The "model" compounds from I and II and benzoylchloride are also obtained. The synthesized polyketones have intrinsic viscosity 0.36–0.84 deciliter/gram (25 °C, H_2SO_4). They do not dissolve in most organic solvents, dissipate in H_2SO_4. It is determined that the polyketones containing the aromatic links are more stable than those involving aliphatic and azo group. The temperature of 10% and 50% mass loss is 150–250 °C and 270–540 °C for these polyketones.

The polyketones, possessing intrinsic viscosity 0.76–1.18 deciliter/gram and badly dissolving in organic solvents, can be produced by Friedel-Crafts polycondensation of diarylidenecyclopentanone or diarylidene-cyclohexanone, chlorides of aromatic or aliphatic diacids, or azodibenzoylchlorides. The temperature of 10% mass loss is 190–300 °C. The in polyketones have the absorption band at wavelength 240–350 nm in ultraviolet spectra (visible range). [357].

The metal-containing polyketones, which do not dissolve in most of organic solvents and easily dissipate in proton dissolvents, have been produced by the reaction of 2,6[bis (2-ferrocenyl) methylene] cyclohexane with chlorides of dicarboxylic acids [358]. The intrinsic viscosity of metal-containing polyketones is 0.29–0.52 deciliter/gram.

The Refs. [359, 360] report on different syntheses of isomeric aromatic polyketones. Three isomeric aromatic polyketones, containing units of 2-trifluoromethyl- and 2, 2{}-dimetoxybisphenylene were synthesized in [326] by means of direct electrophylic aromatic acylated polycondensation of monomers. Two isomers of polyketone of structure "head-to tail" and "head-to head" contain the links of 2-trifrluoromethyl-4,4'{}-bisphenylene and 2,2{}-dimethoxy-5,5{}-bisphenylene.

There exist several reports on syntheses of polyarylketones, containing bisphthalasinone and methylene [361, 362], naphthalene [363–367] links; containing sulfonic groups [368], carboxyl group in side chain [369], fluorine [370–372] and on the basis of carbon monoxide and styrol or n-sthyl-stirol [373]. It is shown that methylene and bisphthalasinone links in main chain of polyketones are responsible for its good solubility in m-cresol, chloroform. The links of bisphthalasinone improve thermal property of polymer.

The synthesis of temperature-resistant polyketone has been carried out in [362] by means of polycondensation of 4-(3-chlor-4-oxyphenyl)-2, 3-phthaloasine-1-one with 4,4'-difluorobisphenylketone.

The polymer is soluble in chloroform, N-methylpirrolydone, nitrobenzene and tetra chloroethane, its glass transition temperature is 267 °C.

The aromatic polyketones, containing 1,4-naphthalene links were produced in Ref. [363] by the reaction of nucleophylic substitution of 1-chlor-4-(4'-chlorbenzoyl)naphthalene with 1) 1,4-hydroquinone, 2) 4,4'-isopropylidenediphenol, 3) phenolphthalein, 4) 4-(4'-hydroxyphenyl)(2H)-phthalasine-1-one, respectively. All polymers are amorphous and dissipate in some organic solvents. The polymers have good thermo-stability and the high glassing temperatures.

Fluorine containing polyarylketone was synthesized on the basis of 2, 3, 4, 5, 6-pentaflourbenzoylbisphenyl esters in Ref. [370]. The polymers possesses good mechanical and dielectric properties, has impact strength, solubility and tolerance for thermooxidative destruction.

The just-produced polyketones are stabilized by treating them with acetic acid solution of inorganic phosphate [371].

The simple fluorine containing polyarylesterketones was obtained in Ref. [372] on the basis of bisphenol AF and 4, 4'difluorobenzophenone. The polymer has the glass transition temperature 163 °C and temperature of 5% mass loss 515 °C, dielectric constant 1.69 at 1 MHz, is soluble well in organic solvents (tetrahydrofurane, dimethylacetamide and others).

The Refs. [374, 375] are devoted to the synthesis of polyesterketones with lateral methyl groups.

The polyarylenesterketones have been produced by the reaction of nucleophylic substitution of 4,4{}-difluorobenzophenone with hydroquinone [374]. The synthesis is held in sulfolane in the presence of anhydrous the K_2CO_3. The increasing content of methylhydroquinone links in polymers leads to increasing of glass transition temperature and lowering of crystallinity degree, melting temperature and activation energy.

The methyl-substituted polyarylesterketones have been produced in Ref. [375] be means of electrophylic polymerization of 4,4{}bis-(0-methylphenoxy)bisphenylketone or 1,4-bis(4-(methylphenoxy)benzoyl)benzene with terephthaloyl or isophthaloyl chloride in 1,2-dichloroethane in the presence of dimethylformamides and $AlCl_3$. The polymers have glassing temperature and melting point 150–170 °C and 175–254 °C.

The simple aromatic polyesterketones, which are of interest as constructional plastics and film materials, capable of operating within the long time at 200 °C, have been synthesized [376–381] by means of polynitrosubstitution reaction of 1,1-dichlor-2, 2-di(4-nitrophenyl)-ethylene and 4,4'dinitrobenzophenone with aromatic bisphenols.

The simple esterketone oligomer can be synthesized [378] also by polycondensation of aromatic diol with halogen-containing benzophenone at 150–250 °C in organic solvent in the presence of alkali metal compound as catalyst and water.

The particles of polyesterketone have a diameter ≤ 50 micrometers, intrinsic viscosity 0.5–2 deciliter/gram (35 °C, the ratio n-chlorphenol: phenol is 90:10).

The simple polyesterketone containing lateral side cyano-groups, possessing the glass transition temperature in range 161–179 °C, has been produced [379] by low-temperature polycondensation of mix 2,6-phenoxybenzonitrile and 4,4'-bisphenoxybenzophenone with terephthaloylchloride in 1,2-dichloroethane.

The simple polyarylesterketones, containing the carboxyl pendent groups [380] and flexible segments of oxyethylene [381] have been synthesized either.

Russian and foreign scientists work on synthesis of copolymers [382–393] and block copolymers [394–399] of aromatic polyketones.

So, the high-molecular polyketones, having amorphous structure and high values of glassing temperature, have been produced by copolymerizing through the mechanism of aromatic combination of 5,5-bis(4-chlorbemzoyl)-2,2-dimethoxybisphenyl and 5,5-bis(3-chlorbenzoyl)-2,2-dimetoxybishenyl with help of nickel complexes [382].

Terpolymers on the basis of 4, 4'bisphenoxybisphenylsulfone, 4,4'bisphenoxybenzophenone and terephthaloylchloride were produced in Ref. [383] by low-temperature polycondensation. The reaction was lead in solution of 1, 2-dichloroethane in the presence of AlCl$_3$ and N-methyl-2-pyrrolidone. It has been found that with increasing content of links of 4, 4'bisphenoxybisphenylsulfone in copolymer their glassing and dissipation temperatures increase, but melting point and temperature of crystallization decrease.

The new copolyesterketones have been produced in Ref. [384] also from 4,4'difluorobenzophenone, 2,2{},3,3{},6,6{}-hexaphenyl-4,4'bisphenyl-1,1{}-diol and hydroquinone by the copolycondensation in solution (sulfolane being the dissolvent) in the presence of bases (Na_2CO_3, K_2CO_3). The synthesized copolyesterketones possess solubility, high thermal stability; have the good breaking strength and good gas-separating ability in relation to CO_2/N_2 and O_2/N_2.

The random copolymers of polyarylesterketones were produced by means of nucleophylic substitution [385] the basis of bisphenol A and carding bisphenols (in particular, phenolphthalein).

The simple copolyesterketones (copolyesterketonearylates) are synthesized [386] and the method for production of copolymers of polyarylates and polycarbonate with polyesterketones is patented [387]. The reaction is held in dipolar aprotone dissolvent in the presence of interface catalyst (hexaalkylguanidinehalogenide).

The statistical copolymers of polyarylesterketones, involving naphthalene cycle in the main chain, can be produced [388] by low-temperature polycondensation of bisphenyloxide, 4,4'{} bis(β-naphtoxy)benzophenone with chloranhydrides of aromatic bicarbonic acids – terephthaloylchloride and isophthaloylchloride (I) in the presence of catalytical system $AlCl_3$/N–methylpirrolydone/$ClCH_2CH_2Cl$ (copolymers are characterized by improved thermo- and chemical stability), and also by the reaction of hydroquinone with 1,4-bis(4,4'flourobenzoyl)naphthalene (II) in the presence of sodium and potassium carbonates in bisphenylsulfone [389].

The glass transition temperature of polyarylesterketones is going up, and melting point and temperature of the beginning of the destructions are down with increasing concentration of links of 1,4-naphthalene in main chain of copolymers.

The polyarylesterketone copolymers, containing lateral cyano-groups, can be synthesized [390] on the basis of bisphenyl oxide, 2,6-bisphenoxybenzonitrile (I) and terephthaloylchloride in the presence of $AlCl_3$, employing 1,2-dichloroethane as the dissolvent, N-mthyl-2-pyrrolidon as the Lewis base. With increasing concentration of links of I the crystallinity degree and the melting point of copolymers decrease while the glass transition temperature increase. The temperature of 5%of copolymers mass loss is more than 514 °C (N_2).

Copolymers, containing 30–40 weight % of links of I, possess higher therm0-stability (350 ± 10 °C), good tolerance for action of alkalis, bases and organic solvents.

The copolymers of polyarylesterketones, containing lateral methyl groups, can be produced by low-temperature polycondensation of 2,2{}-dimcthyl-4,4{}-bisphenoxybisphenyleneketone (I) or 1,4-[4-(2-methylphe-noxy) benzoyl] benzene (II) and bisphenylester, terephthaloylchloride [391]. The synthesis is held in 1, 2-dichloroethane in the presence of di-methylformamides and $AlCl_3$ as catalyst. With increasing concentration of links of I or II in copolymers the glass transition temperature increases, and the melting point and the crystallinity degree decreases.

The copolymers of polyarylketones, containing units of naphthalene-sulfonic acid in lateral links, are use in manufacturing of proton-exchange membranes [392].

The polymeric system, suitable as proton-exchanged membrane in fuel cells, was developed in Ref. [393] on the basis of polyarylenesterketones.

The polyarylesterketones are firstly treated with metasulfonic acid within 12 h at 45 °C up to sulfur content of 1.2%, and then sulfonize with oleum at 45 °C up to sulfur content 5% (the degree of sulfonation 51%).

The high-molecular block-copolyarylenesterketones have been syn-thesized on the basis of 4,4'difluorobenzophenone and number of bisphe-nols by means of reaction of nucleophylic substitution of activated aryl-halogenide in dimethylacetamide (in the presence of potassium carbonate) [394, 395]. It has been identified the cutback of molecular weight of poly-mer when using bisphenolate of 4,4'(isopropylidene)bisphenol.

The block-copolyesterketones are synthesized [396] on the basis of di-chloranhydride-1, 1'dichlor-2,2'-bis(n-carboxyphenyl)ethylene and vari-ous dioxy-compounds with complex of worthy properties. It is possible to create, on the basis of obtained polymers, the constructive and film materials possessing lower combustibility and high insulating properties. Some synthesized materials are promising as modifiers of high-density polyethylene.

The block copolymers, containing units of polyarylesterketones and segments of thermotropic liquid-crystal polyesters of different length, are synthesized by means of high-temperature polycondensation in solution [397, 398]; the kinetics, thermal and liquid-crystal properties of block-copolymers have been studied.

The polyarylenesterketones have been obtained [399], which contain blocks of polyarylesterketones, block of triphenylphosphite oxide and those of binding agent into the cycloaliphatic structures. The film materi-als on their basis are used as polymer binder in thermal control coatings.

The Refs. [400–402] informed on production of polyester-α-diketones using different initial compounds.

Non-asymmetrical polyester-α-diketones are synthesized [400] on the basis of 4-flouru-4' (*n*-fluorophenylglyoxalyl) benzophenone. Obtained polymers are amorphous materials with glassing temperature 162–235 °C and have the temperatures of 10% mass loss in range of 462–523 °C.

The polyester-α-diketones synthesized in [401] on the basis of 2, 2-bis [4-(4-fluorophenylglyoxalyl) phenyl] hexafluoropropane are soluble polymers. They exhibit the solubility in dimethylformamide, dimethylacetamide, N-methylpirrolydone, tetrahydrofurane and chloroform. The values of glassing temperature lie in interval 182–216 °C, and temperatures of 10% mass loss are within 485–536 °C and 534–556 °C in air and argon, respectively.

The number of new polyester-α-diketones in the form of homopolymers and copolymers has been produced in Ref. [402]. Polyester-α-diketones have the amorphous structure, include α-diketone, α-hydroxyketone and ester groups, they are soluble in organic solvents. The films with lengthening of 5–87%, solidity and module under tension 54–83 MPa and 1.6–3 GPa, respectively, are produced by casting from chloroform solution.

The quinoxalline polymers, possessing higher glassing and softening temperatures and better mechanical properties compared to initial polymers, have been produced by means of interaction of polyester-α-diketones with α-phenylenediamine at 23 °C in m-cresol. Polyquinoxalines are amorphous polymers, soluble in chlorinated, amide and phenol dissolvents with η_{lim} = 0.4–0.6 deciliter/gram (25 °C, in N-methylpirrolydone of 0.5 gram/deciliter).

It is reported [403] on liquid-crystal polyenamineketone and on mix of liquid-crystal thermotropic copolyester with polyketone [404]. The mix is prepared by melts mixing. It has been found that the blending agents are partially compatible; the mixes reveal two glass transition temperatures.

The transitions of polyketones into complex polyesters are possible.

So, when conducting the reaction at 65–85 °C in water, alcohol, ester, acetonitrile, the polyketone particles of size 0.01–100 micrometers transform the polyesters by the oxidation under the peroxide agents: peroxybenzoic, m-chloroperoxybenzoic, peroxyacetic, and triflouruperoxyacetic, monoperoxyphthalic, and monoperoxymaleine acids, combinations of H_2O_2 and urea or arsenic acid [405, 406].

For increasing of basic physical-mechanical characteristics and reprocessing, in particular of solubility, the synthesis of aromatic polyketones is lead through the stages of formation of oligomers with end functional groups accompanied by the production of block-copolyketones or through the one-stage copolycondensation initial monomers with production of co-polyketones.

So, unsaturated oligeketons are produced [407, 408] by the condensation of aromatic esters (for example, bisphenyl ester, 4,4′bisphenoxybisphenyl ester and others) with maleic anhydride in the presence catalyst AlCl$_3$.

The oligoketones with end amino groups can be produced [409] on the basis of dichloranhydride of aromatic dicarboxylic acid, aromatic carbo-hydrate and telogen (N-acylanylode) according Friedel-Crafts reaction in organic dissolvent.

The oligoketones can be produced [410] by polycondensation of bi-sphenyloxide and 4-fluorobenzoylchloride in solution, in the presence of AlCl$_3$. It has been found that the structure of synthesized oligomer is crys-talline.

The synthesis of oligoketones, containing phthaloyl links, can be im-plemented [411] by the Friedel-Craft reaction of acylation.

The cyclical oligomers of phenolphthalein of polyarylenestersulfonek-etone can be produced [412] by cyclical depolymerization of correspond-ing polymers in dipolaraprotic solvent (dimethylformamide), dimethylac-etate in the presence of CsF as catalyst.

The aromatic oligoesterketones can be produced by means of inter-action between 4,4′-dichlorbisphenylketone and 1,1-dichlor-2,2-di(3,5-dibrom-n-oxyphenyl)ethylene in aprotic dipolar dissolvent (dimethylsulf-oxide) at 140 °C in inert gas [413]. The copolyesterketones of increased thermo-stability, heat- and fire-resistance can be synthesized on the basis of obtained oligomers.

The works on production of oligoketones and synthesis of aromatic polyketones on their basis are held in Kh.M. Berbekov Kabardino-Balkari-an State University.

So, availing the method of high-temperature polycondensation in en-vironment of dimethylsulfoxide, the oligoketones are produced with edge hydroxyl groups with the degree of condensation 1–20 from bisphenols (Diane or phenolphthalein) and dichlorbenzophenone. The condensation on the second stage is carried out at room temperature in 1,2-dichloroeth-ane in the presence of HCl as acceptor and triethylamine as catalyst with introduction of the diacyldichloride of 1,1-dichlor-2,2-(n-carboxyphenyl) ethylene into the reaction. The reduced viscosity of copolyketones lies within 0.78–3.9 deciliter/gram for polymers on the basis of Diane oligoketones and 0.50–0.85 deciliter/gram for polymers on the basis of phenol-phthalein oligoketones. The mix of the tetrachloroethane and phenol in molar ratio 1:1 at 23 °C was taken as a solvent [414].

The aromatic block-copolyketones have been synthesized by means of polycondensation of oligoketones of different composition and structure with end OH-groups on the basis of 4,4'-dichlorbenzophenone and dichloranhydrides mixed with iso-and terephthalic acids in environment of 1,2-dichloroethane [415–418].

The synthesis is lead via the method of acceptor-catalyst polycondensation. The double excess of triethylamine in respect to oligoketones is used as acceptor-catalyst. Obtained block-copolyketones possess good solubility in chlorinated organic solvents and can be used as temperature-resistant high-strength durable constructional materials.

7.7 AROMATIC POLYESTERSULFONE KETONES

Along with widely used polymer materials of constructional assignment such as polysulfones, polyarylates, polyaryleneketones et cetera, the number of researches has recently, within the last decades, appeared which deal with production and study of properties of polyestersulfoneketones [290, 419–433]. The advantage of given polymeric materials are that these simultaneously combine properties of both polysulfones and polyarylenek-etones and it allows one to exclude some or the other disadvantages of two classes of polymers. It is known, that high concentration of ketone groups results in greater T_{glass} and T_{flow} respectively, or in better heat stability. On the other hand, greater concentration of ketone groups results in worsening of reprocessibility of polyarylenesterketones. That is why the sulfonation of the latter has been performed. The solubility increases with increasing of the degree of sulfonation. Polyestersulfoneketones gain solubility in dichloroethane, chloroform, dimethylformamide. The combination of elementary links of polysulfone with elementary links of polyesterketone also improves the fluidity of composition during extrusion.

The Refs. [424–433] report on various methods of production of poly-estersulfoneketones.

The 4,4{}-bis(phenoxy)bisphenylsulfone has been synthesized by means of reaction of phenol with bis-(4-chlorphenyl)sulfone and the low-temperature polycondensation of the product has been carried out in the melt with tere- and isophthaloylchloride resulting in the formation of polyarylenestersulfonesterketoneketones. The polycondensation is conducted in 1,2-dichlorethane at presence of AlCl$_3$ and N-methylpirrolydone. Compared to polyesterketoneketones synthesized polymers have greater temperatures of glassing and decomposition and lower melting point and also possess higher thermal resistance [424].

The block-copolymers have been created by means of polycondensation of low-molecular polyesterketone containing the remnants of chloranhydride of carbonic acid and 4,4{}-bisphenoxybisphenylsulfone (I) as end groups. The increase in concentration of component I result in greater glassing temperature and lower melting point of block-copolymers.

The block-copolymers containing 32.63–40.7% of component I have glassing temperature and melting point respectively 185–193°C and 322–346°C, durability and modulus in tension respectively 86.6–84.2 MPa and 3.1–3.4 GPa and tensile elongation 18.5–20.3%.

Block-copolymers have good thermal properties and could be easily reprocessed in melt [425].

Polyestersulfoneketones have been produced be means of electrophylic Friedel-Crafts acylation in the presence of dimethylformamides and anhydrous AlCl$_3$ in environment of 1,2-dichlorethane on the basis of simple 2- and 3-methylbisphenyl esters and 4,4{}-bis(4-chloroformylpheoxy) bisphenylsulfone. The copolymer has the molecular weight of 57,000–71,000, its temperatures of glassing and decomposition are 160.5–167.0°C and > 450°C, respectively, the coke end is 52–57% (N$_2$). Copolyestersulfoneketones are well soluble in chloroform and polar dissolvents (dimethylformamide and others) and form transparent and elastic films [426].

Some papers of Chinese researchers are devoted to the synthesis and study of properties of copolymers of arylenestersulfones and esterketones [427–429].

The block-copolyesters based on oligoesterketones and oligoestersulfones of various degree of condensation have been produced in Ref. [430] by means of acceptor-catalyst polycondensation; the products contain simple and complex ester bonds.

Polyestersulfoneketones possessing lower glassing temperatures and excellent solubility have been synthesized in Ref. [431] by means of low-temperature polycondensation of 2, 2{}, 6, 6{}-tetramethylbisphenoxybisphenylsulfone, iso-and terephthaloylchloride in the presence of AlCl$_3$ and N-methylpirrolydone in environment of 1, 2-dichlorethane. The increasing concentration of iso-and terephthaloylchloride results in greater glassing temperatures of statistical copolymers.

The manufacture of ultrafiltration membranes is possible on the basis of mix polysulfone-polyesterketone [432, 433]. Transparent membranes obtained by means of cast from solution have good mechanical properties in both dry and hydrated state and keep analogous mechanical properties after exposition in water (for 24 h at 80 °C). The maximal conductivity of membranes at 23 °C is 4.2 10^{-2} Sm/cm while it increases to 0.11 Sm/cm at 80 °C.

In spite of a number of investigations in the area of synthesis and characterization of polyestersulfoneketones, there are no data in literature on the polyestersulfoneketones of block-composition based on bisphenylolpropane (or phenolphthalein), dichlorbisphenylsulfone and dichlorbenzophenone, terephthaloyl-bis (n-oxybenzoylchloride).

Russian references totally lack any researches on polyestersulfoneketones, not speaking on the production of such polymers in our country. Accounting for this, we have studied the regularities of the synthesis and produced block-copolyestersulfoneketones of some valuable properties [423]. The main structure elements of the polymers are rigid and extremely thermo-stable phenylene groups and flexible, providing for the thermoplastic reprocessibility, ester, sulfone and isopropylidene bridges.

Synthesized, within given investigation, polyestersulfoneketones on the basis of phthalic acid; polyarylates on the basis of 3,5-dibromine-n-oxybenzoic acid and phthalic acids and copolyesters on the basis of terephthaloyl-bis(n-oxybenzoic) acid are of interest as heat-resistant and film materials which can find application in electronic, radioelectronic, avia, automobile, chemical industries and electrotechnique as thermo-stable constructional and electroisolation materials as well as for the protection of the equipment and devices from the influence of aggressive media.

KEYWORDS

- **Monomers**
- **n-oxybenzoic acid**
- **Polycondensation**
- **Polyesters oligoesters**
- **Synthesis**
- **Tere- and isophthalic acids**

REFERENCES

1. Sokolov, L. V. (1976). *The synthesis of polymers: polycondensation method.* Moscow Himia, 332p [in Russian].
2. Morgan, P. U. (1970). *Polycondensation processes of polymers synthesis.* Leningrad Himia, 448p [In Russian]
3. Korshak, V. V., & Vinogradova, S. V. (1968). *Equilibrium polycondensation.* Moscow: Himia, 441p [in Russian].
4. Korshak, V. V., & Vinogradova, S.V. (1972). *Non-equilibrium polycondensation.* Moscow Nauka, 696p [in Russian].
5. Sokolov, L. B. (1979). *The grounds for the polymers synthesis by polycondensation method.* Moscow Himia, 264p [in Russian].
6. Korshak, V. V., & Kozyreva, N. M. (1979). Uspehi himii, *48(1),* 5–29.
7. Encyclopedia of polymers (1977). Moscow Sovetskaia enciklopedia, *3,* 126–138 [in Russian].
8. Korshak, V. V., & Vinogradova, S. V. (1964). *Polyarylates.* Moscow Nauka, 68p [in Russian].
9. Askadskii, A. A. (1969). *Physicochemistry of polyarylates.* Moscow Himia, 211p [in Russian].
10. Korshak, V. V., & Vinogradova, S. V. (1958). *Heterochain polyesters.* Moscow Academy of Sciences of USSR, 403c [in Russian].
11. Lee, G., Stoffi, D., & Neville, K. (1972). *Novel linear polymers.* Moscow Himia, 280p [in Russian].
12. Sukhareva, L. A. (1987). *Polyester covers: structure and properties.* Moscow Himia, 192 p. [in Russian].
13. Didrusco, G., & Valvaszori, A. (1982). Prospettive nel campo bei tecnopolimeri *Tecnopolime resine, (5)* 27–30.
14. Abramov, V. V., Zharkova, N. G., & Baranova, N. S. (1984). Abstracts of the All-Union conference *"Exploiting properties of constructional polymer materials"* Nalchik, 5 [in Russian]
15. Tebbat (1975). Tom. Engineering plastics: Wonder materials of expensive polymer plauthings, *Eur. Chem. News. 27,* 707.
16. Stoenesou, F. A. (1981). *Tehnopolimeri, Rev. Chem., 32(8).* 735–759.
17. Nevskii, L. B., Gerasimov, V. D., & Naumov, V. S. (1984). In Abstracts of the All-Union conference *"Exploiting properties of constructional polymer materials"* Nalchik, 3 [in Russian].
18. Mori Hisao (1975). *Jap. Plast, 26(8),* 23–29.
19. Karis, T., Siemens, R., Volksen, W., & Economy, J. (1987). *Melt processing of the phbahomopolymer* In Abstracts of the 194-th ACS National Meeting of American Chemical Society. New Orleans Los Angeles, August 30–September 4, 1987. Washington DC, 335–336.
20. Buller, K. (1984). *Heat-and thermostable polymers.* Moscow Himia, 343p [in Russian].
21. Crossland, B., Knight, G., & Wright, W. (1986). The thermal degradation of some polymers based upon p-hydroxybenzoic acid, *Brit Polymers. J 18(6),* 371–375.

22. George, E., & Porter, R. (1988). Depression of the crystalnematic phase transition in thermotropic liquid crystal copolyesters, *J. Polym. Sci., 26(1),* 83–90.
23. Yoshimura, T., & Nakamura, M. Wholly aromatic polyester, US Patent 4609720. Publ. 02.09.86.
24. Ueno, S., Sugimoto, H., & Haiacu, K. Method for producing polyarylates, Japan Patent Application 59–120626. Publ. 12.07.84.
25. Ueno, S., Sugimoto, H., & Haiacu, K. Method for producing aromatic polyesters, Japan Patent Application 59–207924 Publ. 26.11.84.
26. Yu Michael C. Polyarylate formation by ester inters change reaction using gamma lactones as diluents, US Patent 4533720.
27. Higashi, F., & Mashimo, T. (1986). Direct polycondensation of hydroxybenzoic acids with thionylchloride in pyridine, *J Polym. Sci. Polym. Chem. Ed. 24(7),* 177–1720.
28. Bykov, V. V., Tyuneva, G. A., Trufanov, A. N. et al. (1986). Izvestia Vuzov. Khimi Him. *Technol, 29(12),* 20–22.
29. Tedzaki, K., & Hiroaka. Method for producing polymers of oxybenzoic acid, Japan Patent 48–23677 Publ. 16.07.73.
30. Process for preparation of oxybenzoyl polymer US Patent 3790528. Publ. 05.02.74.
31. Sima Takeo, Yamasiro Saiti, & Inada Hiroo, Method for producing polyesters of n-oxybenzoic acid Japan Patent 48–37-37355. Publ. 10.11.73.
32. Sakano Tsutomu, Miesi Takehiro. *Polyester fiber* Japan Patent Application, 59–199815. Publ. 13.11.84.
33. Higashi Fukuji & Yamada Yukiharu. Direct polycondensation of hydroxybenzoic acids with bisphenyl chlorophosphate in the presence of esters, *J. Polym. Sci.: Polym. Chem. Ed.*
34. Adxuma Fukudzi. Method for producing complex polyesters, Japan Patent Application 60–60133 Publ. 06.04.85.
35. Chivers, R. A., Blackwell, J., & Gutierrez, G. A. (1985). X-ray Studies of the structure of HBA/HNA copolyesters, In Proceedings of the 2-nd Symposium *"Div. Polym. Chem. Polym. Liq. Cryst."* Washington DC-New York–London, 153–166.
36. Sugijama, H., Lewis, D., & White, J. Structural characteristics, rheological properties, extrusion and melt spinning of 60/40 poly(hydroxybenzoic acidcoethylene terephtalate).
37. Blackwell, J., Dutierrez, G., & Chivers, R. (1985). X-ray studies of thermotropic copolyesters, In Proceedings of the 2nd Symposium *"Div. Polym. Chem. Polym. Liq. Cryst."* Washington (DC)–New York–London, 167–181.
38. Windle, A., & Viney, C. (1985). Golombok R. Molecular correlation in thermotropic copolyesters, *Faraday Discuss. Chem. Soc., 79*–55–72.
39. Calundann Gordon, W. Meet processable thermotropic wholly aromatic polyester containing polybenzoyl units, US Patent 4067852.
40. Morinaga Den, Inada Hiroo, & Kuratsudzi Takatozi. Method for producing thermostable aromatic polyesters, Japan Patent Application, 52–121626 Publ. 13.10.77.
41. Sugimoto Hiroaki & Hanabata Makoto. Method for producing aromatic copolyesters, Japan Patent Application 58–40317 Publ. 09.03.83.
42. Dicke Hans-Rudolf & Kauth Hermann. Thermotrope aromatische Polyester mit hoher steifigkeit, verfahren zu ihrer Herstellung und ihre verwenaung zur Herstel-

lung ven Formkorpern, Filamenten, Fasern und Folien, Germany Patent Application, 3427886.

43. Cottis Steve G. Production of thermally stabilized aromatic polyesters, US Patent 4639504.

44. Matsumoto Tetsuo, Imamura Takayuki, & Kagawa Kipdzi. (1987). *Polyester fiber,* Japan Patent Application 62–133113.

45. Ueno Ryudzo, Masada Kachuiasu, & Hamadzaki Yasuhira. Complex polyesters, Japan Patent Application, 62–68813 Publ. 28.03.87.

46. Tsai Hond Bing, Lee Chyun, & Chang Nien-Shi. (1990). Effect of annealing on the thermal properties of poly (4-hydrohybenzoate-cophenylene isophthalates), *Macromol Chem., 191(6),* 1301–1309.

47. Paul, K. T. (1986). Fire resistance of synthetic furniture. Detection methods, *Fire and Materials, 10(1),* 29–39.

48. Iosida Tamakiho & Aoki Iosihisa (25.09.86). Fire resistant polyester composition, Japan Patent Application, 61–215645.

49. Wang, Y., Wu, D. C., Xie, X. G., & Li, R. X. (1996). Characterization of copoly (p-hydroxybenzoate/bisphenol-A terephthalate) by NMR-spectroscopy, *Polym. J. 28(10),* 896–900.

50. Yerlikaya Zekeriya, Aksoy Serpil, & Bayramli Erdal. (2001). Synthesis and characterization of fully aromatic thermotropic liquid-crystalline copolyesters containing m-hydroxybenzoic acid units, *J. Polym. Sci. A., 39(19),* 3263–3277.

51. Pazzagli Federico, Paci Massimo, Magagnini Pierluigi, Pedretti Ugo, Corno Carlo, Bertolini Guglielmo, & Veracini Carlo (2000). A. Effect of polymerization conditions on the microstructure of a liquid crystalline copolyester, *J. Appl. Polym. Sci., 77(1),* 141–150.

52. Aromatic liquid-crystalline polyester solution composition (2005). US Patent 6838546. International Patent Catalogue C 08 J 3/11, C 08 G 63/19.

53. Yerlikaya Zekeriya, Aksoy Serpil, & Bayramli Erdal. (2002). Synthesis and melt spinning of fully aromatic thermotropic liquid crystalline copolyesters containing m-hydroxybenzoic acid units, *J. Appl. Polym. Sci., 85(12),* 2580–2587.

54. Wang Yu-Zhang, Cheng Xiao-Ting, & Tang Xu-Dong (2002). Synthesis characterization, and thermal properties of phosphorus-containing, wholly aromatic thermotropic copolyesters, *J. Appl. Polym. Sci., 86(5),* 1278–1284.

55. Liquid-crystalline polyester production method (2006). US Patent 7005497. International Patent Catalogue C 08 G 63/00.

56. Wang Jiu-fen, Zhang Na, & Li Cheng-Jie. (2005). Synthesis and study of thermotropic liquid-crystalline copolyester PABA/ABPA/TPA, *Polym. Mater. Sci. Technol., 21(1),* 129–132.

57. Method of producing thermotropic liquid crystalline copolyester (2001). Thermotropic liquid crystalline copolyester composition obtained by the same composition, US Patent 6268419, International Patent Catalogue C 08 K 5/51.

58. Hsiue Lin-tee, Ma Chen-chi M., & Tsai Hong-Bing. (1995). Preparation and characterizations of thermotropic copolyesters of p-hydroxybenzoic acid, sebacic acid, and hydroquinone, *J. Appl. Polym. Sci., 56(4),* 471–476.

59. Frich Dan, Goranov Konstantin, Schneggenburger Lizabeth, & Economy James. (1996). Novel high-temperature aromatic copolyester thermosets, synthesis characterization, and physical properties, *Macromolecules, 29(24,)* 7734–7739.

60. Dong Dewen, Ni Yushan, & Chi Zhenguo. (1996). Synthesis and properties of thermotropic liquid-crystalline copolyesters containing bis-(4-oxyphenyl) methanone II Copolyesters from bis-(4-oxyphenyl) methanone, terephthalic acid, n-oxybenzoic acid and resorcene, *Acta Polym. Sin, 2,* 153–158.

61. Teoh, M. M., Liu, S. L., & Chung, T. S. (2005). Effect of pyridazine structure on thin-film polymerization and phase behavior of thermotropic liquid crystalline copolyesters, *J. Polym. Sci. B, 43(16),* 2230–2242.

62. Liquid crystalline polyesters having a surprisingly good combination of a low melting point, a high heat distortion temperature (1999). A low melt viscosity, and a high tensile elongation, US Patent 5969083. International Patent Catalogue C 08 G 63/00.

63. Aromatic complex polyester (2005). US Patent 6890988. International Patent Catalogue C 08 L 5/3477.

64. Process for producing amorphous anisotropic melt-forming polymers having a high degree of stretchability and polymers produced by same (2001). US Patent 6207790. International Patent Catalogue C 08 G 63/00.

65. Process for producing amorphous anisotropic melt-forming polymers having a high degree of stretchability and polymers produced by same (2000). US Patent 6132884. International Patent Catalogue B 32 B 27/06.

66. Process for producing amorphous anisotropic melt-forming polymers having a high degree of stretchability (2001). US Patent 6222000. International Patent Catalogue C 08 G 63/00.

67. He Chaobin, Lu Zhihua, Zhao Lun, & Chung Tai-Shung. (2001). Synthesis and structure of wholly aromatic liquid-crystalline polyesters containing Meta-and ortholinkages, *J. Polym. Sci. A, 39(8),* 1242–1248.

68. Choi Woon-Seop, Padias Anne Buyle, & Hall, H. K. (2000). LCP aromatic polyesters by esterolysis melt polymerization, *J. Polym. Sci. A., 38(19),* 3586–3595.

69. Chung Tai-Shung, Cheng Si-Xue. (2000). Effect of catalysts on thin-film polymerization of thermotropic liquid crystalline copolyester, *J. Polymers. Sci. A., 38(8),* 1257–1269.

70. Collins, T. L. D., Davies, G. R., & Ward, I. M. (2001). The study of dielectric relaxation in ternary wholly aromatic polyesters, *Polymers Adv. Technol., 12(9),* 544–551.

71. Shinn Ted-Hong, Lin Chen-Chong, & Lin David, C. (1995). Studies on co [Poly (ethylene terephthalate-p-oxybenzoate)] thermotropic copolyester, Sequence distribution evaluated from TSC measurements, *Polymers, 36(2),* 283–289.

72. Poli Giovanna, Paci Massimo, Magagnini Pierluigi, Schaffaro Roberto, & La Mantia Francesco, P. (1996). On the use of PET-LCP copolymers as compatibilizers for PET/LCP blends, *Polym. Eng. and Sci. 36(9),* 1244–1255.

73. Chen Yanming (1998). The study of liquid-crystalline copolyesters PHB/PBT, modified with HQ-TRA, *J. Fushun Petrol. Inst., 18(1),* 26–29.

74. Wang Jiu-fen, Zhu Long-Xin, & Huo Hong-Xing. (2003). The method for producing thermotropic liquid-crystalline complex copolyester of polyethyleneterephthalate, *J. Funct. Polymers, 16(2),* 233–237.

75. Liu Yongjian, Jin, Yi, Haishan, Bu., Luise Robert, R., & Bu Jenny. (2001). Quick crystallization of liquid-crystalline copolyesters based on polyethyleneterephthalate, *J. Appl. Polym. Sci., 79(3),* 497–503.

76. Flores, A., Ania, F., & Balta Calleja, F. J. (1997). Novel aspects of microstructure of liquid crystalline copolyesters as studied by microhardness Influence of composition and temperature, *Polym, 38(21),* 5447–5453.

77. Hall, H. K. Jr., Somogyi Arpad, Bojkova Nina, Padias Anne, B., & Elandaloussi El Hadj (2003). MALDI-TOF analysis of all-aromatic polyesters, In PMSE Preprints, Papers presented at the Meeting of the Division of Polymeric Materials Science and Engineering of the American Chemical Society (New Orleans, 2003), *Amer. Chem. Soc., 88,* 139–140.

78. Takahashi Toshisada, Shoji Hirotoshi, Tsuji Masaharu, Sakurai Kensuke, Sano Hirofumi, & Xiao Changfa. (2000). The structure and stretchability in axial direction of fibers from mixes of liquid-crystalline all-aromatic copolyesters with polyethylene-terephthalate, *Fiber, 56(3),* 135–144.

79. Watanabe Junji, Yuaing Liu, Tuchiya Hitoshi, & Takezoe Hideo. (2000). Polar liquid crystals formed from polar rigid-rod polyester based on hydroxybenzoic acid and hydroxynaphthoic acid, *Mol Cryst and Liq. Cryst. Sci. and Technol. A, 346,* 9–18.

80. Juttner, G., Menning, G., & Nguyen, T. N. (2000). Elastizitatsmodul und Schichten-morphologie von spritzgegossenen LCP-Platten, Kautsch and Gummi Kunstst B, 53S 408–414.

81. Bharadwaj Rishikesh & Boyd Richard, H. (1999). Chain dynamics in the nematic melt of an aromatic liquid crystalline copolyester, A molecular dynamics simulation study, *J Chem Phys, 1(20),* 10203–10211.

82. Wang, Y., Wu, D. C., Xie, X. G., & Li, R. X. (1996). Characterization of copoly (p-hydroxybenzoate/bisphenol-A terephthalate) by NMR-spectroscopy, *Polym. J. 28(10),* 896–900.

83. Ishaq M., Blackwell J., & Chvalun S. N. (1996). Molecular modeling of the structure of the copolyester prepared from p-hydroxybenzoic acid, bisphenol and terephthalic acid, *Polym, 37(10),* 1765–1774.

84. Cantrell, G. R., McDowell, C. C., Freeman, B. D., & Noel, C. (1999). The influence of annealing on thermal transitions in nematic copolyester, *J. Polym. Sci. B., 37(6),* 505–522.

85. Bi Shuguang, Zhang, Yi, Bu Haishan, Luise Robert, R., & Bu Jenny, Z. (1999). Thermal transition of wholly aromatic thermotropic liquid crystalline copolyester, *J. Polym. Sci. A, 37(20),* 3763–3769.

86. Dreval, V. E., Al-Itavi Kh, I., Kuleznev, V. N., Bondarenko, G. N., & Shklyaruk, B. F. (2004). *Vysokomol. Soed. A, 46(9),* 1519–1526.

87. Tereshin, A. K., Vasilieva, O. V., Avdeev, N. N., Bondarenko, G. N., & Kulichihin, V. G. (2000). *Vysokomol. Soed. A-B, 42(6),* 1009–1015.

88. Yamato Masafumi, Murohashi Ritsuko, Kimura Tsunehisa, & Ito Eiko. (1997). Dielectric β-relaxation in copolymer ethyleneterephthalate-p-hydroxybenzoic acid, *J. Polym. Sci. And Technol, 54(9),* 544–551.

89. Carius Hans-Eckart, Schonhals Andreas, Guigner Delphine, Sterzynski Tomasz, & Brostow Witold (1996). Dielectric and mechanical relaxation in the blends of a polymer liquid crystal with polycarbonate, *Macromolecules, 29(14),* 5017–5025.

90. Tereshin, A. K., Vasilieva, O. V., Bondarenko, G. N., & Kulichihin, V. G. (1995). Influence of interface interaction on rheological behavior of mixes of polyethyleneterephthalate with liquid-crystalline polyester, *In Abstracts of the III Russian symposium on liquid-crystal polymers*, Chernogolovka, 124 [in Russian].

91. Dreval, V. E., Kulichihin, V. G., Frenkin, E. I., & Al-Itavi Kh.I. (2000). *Vysokomol. Soed. A-B, 42(1),* 64–70.

92. Kotomin, S. V., Kulichihin, V. G. (1996). Determination of the flow limit of LQ polyesters with help of method of parallel-plate compression, In Abstracts of the 18-th Symposium of Rheology, Karacharovo, 61 [in Russian].

93. Zhang Guangli, Yan Fengqi, Li Yong, Wang Zhen, Pan Jingqi, & Zhang Hongzhi (1996). The study of liquid-crystalline copolyesters of n-oxybenzoic acid and polyethyleneterephthalate, *Acta polym. Sin. 1,* 77–81.

94. Dreval, V. E., Frenkin, E. I., Al-Itavi Kh, I., & Kotova, E. V. (1999). Some thermophysical characteristics of liquid-crystalline copolyester based on oxybenzoic acid and polyethyleneterephthalate at high pressures/in Abstracts of the IV-th Russian Symposium (involving international participants) *"Liquid crystal polymers"* Moscow 62 [in Russian].

95. Brostow Witold, Faitelson Elena, A., Kamensky Mihail, G., Korkhov Vadim P., & Rodin Yuriy, P. (1999). Orientation of a longitudinal polymer liquid crystal in a constant magnetic field, *Polym. 40(6),* 1441–1449.

96. Dreval, V. E., Hayretdinov, F. N., Kerber, M. L., & Kulichihin, V. G. (1998). *Vysokomol Soed A-B, 40(5),* 853–859.

97. Al-Itavi Kh, I., Frenkin, E. I., Kotova, E. V., Bondarenko, G. N., Shklyaruk, B. F., Kuleznev, V. N., Dreval, V. E., & Antipov, E. M. (2000). Influence of high pressure on structure and thermophysical properties of mixes of polyethyleneterephthalate with liquid-crystalline polymer, In Abstracts of the 2-nd Russian Kargin Symposium *"Chemistry and physics of polymers in the beginning of the 21 century"* Chernogolovka, Part 1–1/13 [in Russian].

98. Garbarczyk, J., & Kamyszek, G. (2000). Influence of magnetic and electric field on the structure of IPP in blends with liquid crystalline polymers, In Abstracts of the 38-th Macromolecular IUPAK Symposium. Warsaw, 3–1195.

99. Dreval, V. E., Frenkin, E. I., Kotova, E. D. (1996). Dependence of the volume form the temperature and pressure for thermotropic LQ-polymers and their mixes with polypropylene, In Abstracts of the 18-th Symposium on Rheology. Karacharovo, 45 [in Russian].

100. Al-Itavi Kh, I., Dreval, V. E., Kuleznev, V. N., Kotova, E. V., & Frenkin, E. I. (2003). *Vysokomol Soed, 45(4),* 641–648.

101. Plotnikova, E. P., Kulichihin, E. P., Mihailova, I. M., & Kerber, M. L. (1996). Rotational and capillary viscometry of melts of mixes of traditional and liquid-crystalline thermoplasts, In Abstracts of the 18-th Symposium on Rheology. Karacharovo, 117 [in Russian].

102. Kotomin, S. V., & Kulichihin, B. G. (1999). Flow limit of melts of liquid-crystal polyesters and their mixtures/in Abstracts of the IV Russian Symposium (involving international participants) "Liquid crystal polymers." Moscow 63 [in Russian].

103. Park Dae Soon, & Kim Seong Hun. (2003). Miscibility study on blend of thermotropic liquid crystalline polymers and polyester, *J. Appl. Polym. Sci, 87(11)* 1842–1851.

104. Bharadwaj Rishikesh, K., & Boyd Richard, H. (1999). Diffusion of low-molecular penetrant into the aromatic polyesters: modeling with method of molecular dynamics, *Polymer, 40(15),* 4229–4236.

105. Luscheikin, G. A., Dreval, V. E., & Kulichihin, V. G. (1998). *Vysokomol. Soed. A-B, 40(9),* 1511–1515.

106. Shumsky, V. F., Getmanchuk, I. P., Rosovitsky, V. F., & Lipatov, Yu.S. (1996). Rheological viscoelastic and mechanical properties of mixes of polymethylmethacrylate with liquid-crystal copolyester filled with wire-like monocrystals, In Abstracts of the 18-th Symposium on Rheology, Karacharovo, p. 115 [in Russian].

107. Liu Yongjian, Jin, Yi, Dai Linsen, Bu Haishan, & Luise Robert, R. (1999). Crystallization and melting behavior of liquid crystalline copolyesters based on poly (ethyleneterephthalate), *J. Polym. Sci. A. 37(3),* 369–377.

108. Abdullaev Kh, M., Tuichiev Sh, T., Kurbanaliev, M. K., & Kulichihin, V. G. (1997). *Vysokomol Soed A-B, 39(6),* 1067–1070.

109. Li Xin-Gui. (1999). Structure of liquid crystalline copolyesters from two acetoxybenzoic acids and polyethyleneterephthalate, *J. Appl. Polym. Sci., 73(14),* 2921–2925.

110. Li Xin-Gui & Huang Mei-Rong (1999). High-resolution thermogravimetry of liquid crystalline copoly (p-oxybenzoateethyleneterephthalate-m-oxybenzoate), *J. Appl. Polym. Sci., 73(14),* 2911–2919.

111. Guo Mingming & Britain William, J. (1998). Structure and properties of naphthalene-containing polyesters. 4. New insight into the relationship of transesterification and miscibility, *Macromolecules, 31(21),* 7166–7171.

112. Li Xin-Gui, Huang Mei-Rong, & Guan Gui-He, Sun Tong. (1996). Glass transition of thermotropic polymers based upon vanillic acid, p-hydroxybenzoic acid, and poly (ethyleneterephthalate), *J. Appl. Polym. Sci., 59(1),* 1–8.

113. Additives and modifiers (1987–1988). *Plast Compound, 4,* pp. 10, 14–16, 18, 20, 24, 26, 28, 30, 32, 34, 36, 38–40, 42–44, 46–51.

114. Sikorski R., Stepien A. (1972). Nienasycone zywice poliestrowe zawierajace cherowiec, Cz. J., Studie problemowe Prnauk inst. technol. Organicz. It worzyw. sztuczn. PWr., *7,* 3–19.

115. Takase, Y., Mitchell, G., & Odajima, A. (1986). Dielectric behavior of rigid chain thermotropic copolyesters, *Polym. Commun. 27(3),* 76–78.

116. Volchek, B. Z., Holmuradov, N. S., Bilibin, A. Yu., & Skorohodov, S. S. (1984). *Vysokomol Soed, A 26(1),* 328–333.

117. Bolotnikova, L. S., Bilibin, A. Yu., Evseev, A. K., Panov, Yu. N., Skorohodov, S. S., & Frenkel, S. I. A. (1983). *Vysokomol Soed. A, 25(10),* 2114–2120.

118. Volchek, B. Z., Holmuradov, N. S., Purkina, A. V., Bilibin, A. Yu., & Skorohodov, S. S. (1984). *Vysokomol Soed A., 27(1),* 80–84.

119. Andreeva, L. N., Beliaeva, E. V., Lavrenko, P. N., Okopova, O. P., Tsvetkov, V. N., Bilibin, A. Yu., & Skorohodov, S. S. (1985). *Vysokomol Soed A, 27(1),* 74–79.

120. Grigoriev, A. N., Andreeva, L. N., Bilibin, A. Yu., Skorohodov, S. S., & Eskin, V. E. (1984). *Vysokomol Soed A, 26(8),* 591–594.

121. Grigoriev, A. N., Andreeva, L. N., Matveeva, G. I., Bilibin, A. Yu., Skorohodov, S. S., & Eskin, V. E. (1985). *Vysokomol Soed B, 27(10),* 758–762.

122. Bolotnikova, L. S., Bilibin, A. Yu., Evseev, A. K., Ivanov, Yu. N., Piraner, O. N., Skorohodov, S. S., & Frenkel, S. Ya. (1985). *Vysokomol Soed A, 27(5),* 1029–1034.

123. Pashkovsky, E. E. (1986). Abstracts of the Thesis for the scientific degree of candidate of physical and mathematical sciences. Leningrad, 19p [in Russian]

124. Grigoriev, A. N., Andreeva, L. N., Volkov, A. Ya., Smirnova, G. S., Skorohodov, S. S., & Eskin, V. E. (1987). *Vysokomol Soed A, 29(6)*, 1158–1161.

125. Grigoriev, A. N., Matveeva, G. I., Piraner, O. N., Lukasov, S. V., Bilibin, A. Yu., Sidorovich, A. V. (1991). *Vysokomol Soed A, 33(6)*, 1301–1305.

126. Liquid crystal order in polymer, Blumshtein, A. Ed. Moscow.

127. Grigoriev, A. N., Matveeva, G. I., Lukasov, S. V., Piraner, O. N., Bilibin, A. Yu., Sidorovich, A. V. (1990). *Vysokomol Soed A-B, 32(5)*, 394–396.

128. Bilibin, A. Yu. (1988). *Vysokomol Soed B, 31(3)*, 163.

129. Kapralova, V. M., Zuev, V. V., Koltsov, A. I., Skorohodov, S. S., & Khachaturov, A. S. (1991). *Vysokomol Soed A, 33(8)*, 1658–1662.

130. Helfund, E. (1971). *J. Chem., Phys., 54(11)*, 4651.

131. Bilibin, A. Yu., Piraner, O. N., Skorohodov, S. S., Volenchik, L. Z., & Kever, E. E. (1990). *Vysokomol Soed A, 32(3)*, 617–623.

132. Matveeva, G. N. (1986). Abstracts of the thesis for the scientific degree of candidate of physical and mathematical sciences, 17 p [in Russian].

133. Volkov, A. Ya., Grigoriev, A. I., Savenkov, A. D., Lukasov, S. V., Zuev, V. V., Sidorovich, A. V., & Skorohodov, S. S. (1994). *Vysokomol Soed B, 36(1)*, 156–159.

134. Andreeva, L. N., Bushin, S. V., Matyshin, A. I., Bezrukova, M. A., Tsvetov, V. N., Bilibin, A. Yu., Skorohodov, S. S. (1990). *Vysokomol Soed A, 32(8)*, 1754–1759.

135. Stepanova, A. R. (1992). Abstracts of the Thesis for the scientific degree of candidate of chemical sciences, Sankt–Petersburg, 24p [in Russian].

136. He Xiao-Hua, Wang Xia-Yu. (2002). Synthesis and properties of thermotropic liquid-crystalline block-copolymers containing links of polyarylate and thermotropic liquid-crystalline copolyester (HTH-6), *Natur. Sci. J. Xiangtan Univ., 23(1)*, 49–52.

137. Wang Jiu-fen, Zhu-xin, Huo Hong-xing (2003). *J. Funct. Polymer, 16(2)*, 233–237.

138. Jo Byung-Wook, Chang Jin-Hae, & Jin Jung-2. (1995). Transesterifications in a polyblend of poly(butylene terephthalate) and a liquid crystalline polyester, *Polym. Eng. and Sci., 35(20)*, 1615–1620.

139. Gomez, M. A., Roman, F., Marco, C., Del Pino, J., & Fatou, J. G. (1997). Relaxations in poly (tetramethylene terephtaloyl-bis-4-oxybenzoate): effect of substitution in the mesogenic unit and in the flexible spacer, *Polymer. 38(21)*, 5307–5311.

140. Bilibin, A. Yu., Shepelevsky, A. A., Savinova, T. E., & Skorohodov, S. S. (1982). Terephthaloyl-bis-n-oxybenzoic acid or its dichloranhydride as monomer for the synthesis of thermotropic liquid-crystalline polymers, USSR Inventor Certificate 792834. International Patent Catalogue C 07 C 63/06, C 08K 5/09.

141. Storozhuk, I. P. (1976). Regularities of the Formation of Poly and Oligoarylenesulfonoxides and Block-Copolymers on their Base. *Thesis for the scientific degree of candidate of chemical sciences*, Moscow, 195p [in Russian].

142. Rigid polysulfones hold at 300F (1965). *Jron Age., 195(15)*, 108–109.

143. High-temperature thermoplastics (1965). *Chem Eng Progr., 61(5)*, 144.

144. Thermoplastic polysulfones strength at high temperatures (1965). *Chem Eng Progr., 72(10)*, 108–110.

145. Polysulfones (1966). *Brit. Plast., 39(3)*, 132–135.

146. Lapshin, V. V. (1967). *Plast. Massy 1*, 74–78.

147. Gonezy, A. A. (1979). Polysulfon-ein Hochwarmebestandiger, transparenter Kunststof, *Kunststoffe Bild., 69(1)*, 12–17.
148. Thornton, E. A. (1968). Polysulfone thermoplastics for engineering, *Plast. Eng.*
149. Moiseev, Yu., V., & Zaikov, G. E. (1979). Chemical stability of polymers in aggressive media, Moscow Himia, 288 p. [in Russian]
150. Thornton, E. A., & Cloxton, H. M. (1968). Polysulfones properties and processing characteristics, *Plastics, 33(364)*, 178–191.
151. Huml, J., & Doupovcova, J. (1970). Polysulfon -nogy druh suntetickych pruskuric, *Plast. Hmoty akanc., 7(4)*, 102–106.
152. Morneau, G. A. (1970). Thermoplastic polyarylenesulfone that can be used at 500 °F, *Mod Plast., 47(1)*, 150–152, 157.
153. Storozhuk, I. P., & Valetsky, P. M. (1978). *Chemistry and Technology of high-molecular compounds, 2*, 127–176.
154. Benson, B. A., Bringer, R. P., & Jogel, H. A. (1967). Polymer 360 a thermoplastic for use at 500 °F presented at SPE Antes, Detroit, Michigan.
155. Jdem, A. (1967). Phenylene thermoplastic for use at 500 °F, *SPE Journal.*
156. Besset, H. D., Fazzari, A. M., & Staub, R. B. (1965). *Plast Technol, 11(9)*, 50.
157. Jaskot, E. S. (1966). *SPE Journal, 22*, 53.
158. Leslie, V. J. (1974). Properties at application des polysulfones, *Rev. Gen. Caontch, 51(3)*, 159–162.
159. Bringer, R. P., & Morneau, G. A. (1969). Polymer 360 a new thermoplastic polysulfone for use at 500 °F, *Appl Polym. Symp, 11*, 189–208.
160. Andree U. (1974). Polyarilsulfon ein ansergewohnliecher Termoplast Kunstst of, *Kunststoffe Bild, 64(11)*, 684.
161. Giorgi, E. O. (1971). Termoplastico de engenharia ideal para as condicoes Brasileiras, *Rev. Guim. Ind., 40(470)*, 16–18.
162. Korshak, V. V., Storozhuk, I. P., & Mikitaev, A. K. (1976). Polysulfones-sulfonyl containing polymers, in Polycondensation processes and polymers, Nalchik, 40–78 [in Russian].
163. Two tondh resistant plastic sthrive in hot environments (1969). Prod. Eng., 40(14), 112.
164. Polysulfonic aromatici (1972). *Mater. Plast ed Elast., 38(12)*, 1043–1044.
165. Rose, J. B. (1974). *Polymer, 15(17)*, 456–465.
166. Rigby, R. B. (1979). Victrex-polyestersulfone, *Plast Panorama Scand, 29(11)*, 10–12.
167. Gonozy, A. A. (1979). Polysulfon-ein hochwarmebeston dider transparenter Kunststoff, *Kunstst offe, Bild, 69(1)*, 12–17.
168. Un nuovo tecnotermoplastico in polifenilsulfone radel (1977). *Mater. Plast ed Elast, 2.* 83–85.
169. Polyestersulfon in der BASF (1982) Palette Gimmi Asbest, Kunststoffe, *Bild, 35(3),* 160–161.
170. Bolotina, L. M., & Chebotarev, V. P. (2003). *Plast Massy, 11*, 3–7.
171. Militskova, A. M., & Artemov, S. V. (1990). Aromatic polysulfones polyester (ester) ketones, polyphenylenoxides and polysulfides of NIITEHIM: Review. Moscow, 1–43 [in Russian].

172. High-durable plastics (1999). Kunststoffe, du "hart im Nehmen" sind Technica (Suisse). *Bild, 48,* 25–26, 16–22, [in German].
173. Kampf Rudolf. The method for producing polymers by means of condensation in melts (polyamides, polysulfones, polyacrylates, etc.). Germany Patent Application 102004034708. International Patent Catalogue C 08 P 85/00.2006.
174. Asueva, L. A. (2010). Aromatic polyesters based of terephthaloyl-bis-(n-oxybenzoic) acid, Thesis for the scientific degree of candidate of chemical sciences. Nalchik: KBSU, 129 p. [in Russian]
175. Japan Patent Application 1256524. (1989).
176. Japan Patent Application 1315421. (1995).
177. Japan Patent Application 211634. (1990).
178. Japan Patent Application 1256525. (1989).
179. Japan Patent Application 12565269. (1989).
180. Macocinschi Doina, Grigoriu Aurelia, & Filip Daniela (2002). Aromatic polyculfones used for decreasing of combustibility, *Eur. Polym. J., 38(5),* 1025–1031.
181. US Patent 6548622. (2003).
182. Synthesis and characterization poly (arylenesulfone) (2002). *J. Polym. Sci. A., 40(4).* 496–510.
183. Germany Patent Application 19926778. (2000).
184. Vologirov, A. K., & Kumysheva Yu.A. (2003). Vestnik KBGU Seria Himicheskih nauk 5 P. 86 [in Russian]
185. Mackinnon Sean M., Bender Timothy P., & Wang Zhi Yuan (2000). Synthesis and properties of polyestersulfones, *J. Polym. Sci. A., 38(1),* 9–17.
186. Khasbulatova, Z. S., Asueva, L. A., & Shustov, G. B. (2009). Polymers on the basis of aromatic oligosulfones, in Proceedings of the X International conference on chemistry and physicochemistry of oligomers. Volgograd, P., 100 [in Russian]
187. Ilyin, V. V., & Bilibin, A. Yu. (2002). Synthesis and properties of multiblock-copolymers consisting of flexible and rigid-link blocks, in Materials of the 3-rd Youth school-conference on organic synthesis. Sankt-Petersburg, 230–231 [in Russian]
188. Germany Patent Application 19907605. (2000).
189. Reuter, Knud, Wollbom Ute, & Pudleiner Heinz. (2000). Transesterification as novel method for the synthesis of block-copolymers of simple polyester-sulfone, in Papers of the 38-th Macromolecular IUPAC Symposium., Warsaw, 34.
190. Zhu Shenmin, Xiao Guyu, & Yan Deyue (2001). Synthesis of aromatic graft copolymers, *J. Polym. Sci. A., 39(17),* 2943–2950.
191. Wu Fangjuan, Song CaishEng., Xie Guangliang, & Liao Guihong. (2007). Synthesis and properties of copolymers of 4.4'-bis-(2-methylphenoxy) bisphenylsulfone, 1,4-bisphenoxybenzebe and terephthaloyl chloride, *Acta Polym. Sin, 12,* 1192–1195.
192. Ye Su-fang, Yang Xiao-hui, Zheng Zhen, Yao Hong-xi, & Wang Ming-jun (2006). The synthesis and characterization of novel aromatic polysulfones polyurethane containing fluorine, 40(7), 1239–1243.
193. Ochiai Bundo, Kuwabara Kei, Nagai Daisuke, &, Miyagawa Toyoharu (2006). Endo Takeshi Synthesis and properties of novel polysulfone bearing exomethylene structure, *Eur. Polym. J, 42(8),* 1934–1938.

194. Bolotina, L. M., & Chebotarev, V. P. (2007). The method for producing the statistical copolymers of polyphenylenesulfidesulfones, RF Patent 2311429, International Patent Catalogue C 08 G 75/20.
195. Kharaev, A. V., Bazheva, R. Ch., Barokova, E. B., Istepanova, O. L., & Chaika, A. A. (2007). Fire-resistant aromatic block-copolymers based on 1, 1-dichlor-2, 2- bis(п-oxyphenyl)ethylele, in Proceedings of the 3-rd Russian Scientific and Practical Conference. Nalchik, 17–21 [in Russian]
196. Saxena Akanksha, Sadhana, R., Rao, V., Lakshmana Ravindran, P. V., & Ninan, K. N. (2005). Synthesis and properties of poly (ester nitrile sulfone) copolymers with pendant methyl groups, *J. Appl Polym. Sci., 97,* 1987–1994.
197. Linares A., & Acosta, J. L. (2004). Structural characterization of polymer blends based on polysulfones, *J. Appl. Polym. Sci, 92(5),* 3030–3039.
198. Ramazanov, G. A., Shahnazarov, R. Z., & Guliev, A. M. (2005). *Russian J. Appl. Chem., 78(10),* 1725–1728.
199. Zhao Qiuxia, & Hanson James, E. (2006). Direct synthesis of poly (arylmethyl sulfone) monodendrons, *Synthesis, 3,* 397–399.
200. Cozan, V., & Avram, E. (2003). Liquid-crystalline polysulfone possessing thermotropic properties, *Eur Polym. J., 39(1),* 107–114.
201. Dass, N. N. (2000). *Indian J. Phys. A., 74(3),* 295–298.
202. Zhang Qiuyu, Xie Gang, Yan Hongxia, Xiao Jun, & Li Yurhang. (2001). The effect of compatibility of polysulfone and thermotropic liquid-crystalline polymer, *J. North-West. Polytechn Univ., 19(2),* 173–176.
203. Magagnini, P. L., Paci, M., La Mantia, F. P., Surkova, I. N., & Vasnev, V. A. (1995). Morphology and rheology of mixes from sulfone and polyester Vectra-A 950, *J Appl. Polym. Sci., 55(3),* 461–480.
204. Garcia M., Eguiazabal, J. L., & Nuzabal, J. (2004). Morphology and mechanical properties of polysulfones modified with liquid-crystalline polymer, *J. Macromol. Sci. B., 43(2),* 489–505.
205. RF Patent Application 93003367/04. (1996).
206. Wang Li-jiang, Jian Xi-gao, Liu Yan-jun, &, Zheng Guo-dong. (2001). Synthesis and characterization of polyarylestersulfoneketone from 1-methyl-4, 5-bis (chlorbenzoyl)-cyclohexane and 4-(4-hydroxyphenyl)-2, 3-phthalasin-1-one, *J. Funct. Polym, 14(1),* 53–56.
207. Lei Wei, &, Cai Ming-Zhong. (2004) Synthesis and properties of block-copolymers of polyesterketoneketone and 4, 4{}-bisphenoxybisphenylsulfone, *J. Appl. Chem., 21(7),* 669–672.
208. Tong Yong-fen, Song Cai-shEng., Wen Hong-li, Chen Lie, &, Liu Xiao-ling. (2005). Synthesis and properties of co polymers of aryl ester sulfones and esteresterketones containing methyl replacers, *Polymers Mater. Sci. Technol, 21(2),* 162–165.
209. Bowen, W., Richard, Doneva Teodora, A., & Yin, II. B. (2000). Membranes made from polysulfone mixed with polyesteresterketone, systematic synthesis and characterization, Program and Abstr Tel Aviv, 266.
210. Zinaida, S., Khasbulatova, Luisa, A., Asueva, Madina, A., Nasurova, Arsen, M., Karayev, & Gennady B. (2006). Shustov Polysulfonesterketones on the oligoester base, their thermo- and chemical resistance. 99–105.

211. Khasbulatova, Z. S., Asueva, L. A., Nasurova, M. A., Kharaev, A. M., & Temiraev, K. B. (2005). Simple oligoesters: properties and application, in Proceedings of the 2-nd Russian Scientific and Practical Conference. Nalchik, 54–57 [in Russian]

212. Khasbulatova, Z. S., Asueva, L. A., Nasurova, M. A., Shustov, G. B., Temiraev, K. B., Kharaeva, R. A., & Asibokova, O. R. (2006). Synthesis and properties of aromatic oligoesters in Materials of International Conference on Organic Chemistry, "Organic Chemistry from Butlerov and Belshtein till nowadays" Sankt-Petersburg, 793–794 [in Russian]

213. Iuске, A. (1990). Polyarylketone (PAEK), *Kunststoffe Bild., 80(10),* 1154–1158, 1063.

214. Khirosi, I. (1983). Polyesterketone Victrex PEEK, *31(6),* 31–36.

215. Teruo, S. (1982). Properties and application of special plastics. Polyesteresterketone, Koge Dzaire. 30(9), 32–34.

216. Hay, I. M., Kemmish, D. I., Landford, I. J., & Rae, A. J. (1984). The structure of crystalline PEEK, *Polym. Commun., 25(6),* 175–179.

217. Andrew, I., Lovinger, & Davis, D. D. (1984). Single crystals of poly (ester-ester-ketone) (PEEK) *Polym. Commun., 25(6),* 322–324.

218. Wolf, M. (1987). Anwendungstechnische Entwicklungen bie polyaromaten, Kunststoffe Bild, *77(6),* 613–616.

219. Schlusselindustrien fur technische Kunststoffe (1987). Plastverarbeiter Bild., 38(5), 46–47, 50.

220. May, R. Jn. (1984). In Proceedings of the 7-th Anme Des Eng. Conf. Kempston, 313–318.

221. Rigby Rhymer B. (1984). Polyesteretperketone PEEK, *Polymer News, 9,* 325–328.

222. Attwood, T. E., Dawson, P. C., & Freeman, I. L. (1979). Synthesis and properties of polyarylesterketones, *Amer Chem Soc Polym. Prepr, 20(1),* 191–194.

223. Kricheldorf, H. R., & Bier, G. (1984). New polymer synthesis 11 Preparation of aromatic poly (ester ketones) from silylated bisphenols, *Polymer, 25(8),* 1151–1156

224. Polyesterketone (1986). *High Polym. Jap., 35(4).* 380.

225. High Heat Resistant Film-Talpa (1986). *Japan Plastics Age, 24(208),* 30.

226. Takao Ia. (1988). Polyestersulfones, polyesterketones, *Koge Dzaire End Mater, 36(12),* 120–121.

227. Takao Ia. (1990). Polyesterketones, *Koge Dzaire End Mater, 38(3),* 107–116.

228. Khasbulatoba, Z. S., Kharaev, A. M., Miritaev, A. K. et al. (1992). *Plast Massy, 3,* 3–7.

229. Hergentother, P. M. (1987). Recent advances in high temperature polymers *Polym. J, 19(1),* 73–83.

230. Method for producing of the aromatic polymer in the presence of inert nonpolar aromatic plastificator, US Patent 4110314.

231. Method for producing polyesters, Germany Patent Application 2731816.

232. Aromatic simple polyesters, GB Patent 1558671.

233. Method for producing polyesters, Germany Patent Application 2749645.

234. Producing of aromatic simple polyesters, GB Patent 1569603.

235. Method for producing aromatic polyesters, GB Patent 1563222.

236. Producing of simple aromatic polyesters containing microscopic inclusions of non-melting compounds, US Patent 4331798.

237. Method for producing aromatic polymers, Japan Patent 57–23396.
238. Wear-resistant, self-lubricating composition, Japan Patent Application 58–109554.
239. Antifriction composition, Japan Patent Application 58–179262.
240. Thermoplastic aromatic polyesterketone, Japan Patent 62–146922.
241. Composition on the basis of aromatic polyarylketones, Japan Patent Application 63–20358.
242. New polyarylketones, US Patent 4731429.
243. Method for producing crystalline aromatic polyesterketone, US Patent 4757126.
244. Method for producing high-molecular simple polyesters, Japan Patent Application 63–95230.
245. Aromatic simple esters and method for producing same, Japan Patent Application 63–20328.
246. Method for producing aromatic simple polyesters, Japan Patent Application 63–20328.
247. All-aromatic copolyester, Japan Patent Application 63–12360.
248. All-aromatic copolyesters, Japan Patent Application 63–15820.
249. Colguhoun, H. M. (1984). Synthesis of polyesterketones in trifluoro methane sulphonic acid, some structure-reactivity relationships, *Amer Chem. Soc Polym. Prepr, 25(2),* 17–18.
250. Polyesterketones, Japan Patent Application 60–144329.
251. Method for producing polyesterketones, Japan Patent Application 61–213219.
252. New polymers and method for producing same, Japan Patent Application 62–11726.
253. Method for producing simple polyesterketones, Japan Patent Application 63–75032.
254. Process for producing aromatic polyesterketones, US Patent 4638944.
255. Method for producing crystalline aromatic simple polyesterketones, Japan Patent Application 62–7730.
256. Method for producing thermoplastic aromatic simple polyesters, Japan Patent Application 62–148524.
257. Method for producing thermoplastic polyesterketones, Japan Patent Application 62–148323.
258. Thermoplastic aromatic simple polyesterketones and method producing same, Japan Patent Application 62–151421.
259. Method for producing polyarylesterketones using catalyst on the basis of sodium carbonate and salt of organic acid, US Patent 4748227.
260. Thermostable polyarylesterketones, Germany Patent Application 37008101.
261. Simple aromatic polyesterketones, Japan Patent Application 63–120731.
262. Impact strength polyarylesterketones, Japan Patent Application 63–120730.
263. Method for producing simple polyarylesterketones in the presence of salts of lanthanides, alkali and alkali-earth metals, US Patent 4774311.
264. Jovu, M., & Marinecsu, G. (1981). Rolicetoeteri. Produce de policondensaze ale 4, 4–dihidroxibenzofenonei cu compusi bisclorometilate aromatic, *Rev. Chim. 32(12),* 1151–1158.
265. Sankaran, V., & Marvel, C. S. (1979). Polyaromatic ester-ketone-sulfones containing 1, 3-butadiene units, *J. Polymer Sci. Polymer Chem. Ed. 17(12),* 3943–3957.
266. Method for producing aromatic polyesterketones, Japan Patent Application 60–101119.

267. Method for producing aromatic polyester ketones and polythioester ketones, US Patent 4661581.
268. Method for producing aromatic polyester ketones, Germany Patent Application 3416446.
269. Uncrosslinked-linked thermoplastic reprocessible polyester ketone and method for its production, Germany Patent Application 3416445 A.
270. Litter, M. J., & Marvel, C. S. (1986). Polyaromatic esterketones and polyaromatic ester-ketone sulfonamides from 4-phenoxy-benzoyl ester, *J. Polym. Sci.Polym. Chem. Ed., 23(8)*, 2205.
271. Method for producing aromatic simple poly (thio) esterketone, Japan Patent Application 61–221228.
272. Method for producing aromatic simple poly (thio) esterketone, Japan Patent Application 61–221229.
273. Method for producing polyarylketone involving treatment with the diluents, US Patent 4665151.
274. Copolyesterketones, US Patent 4704448.
275. Method for producing aromatic poly (thio) esterketones, Japan Patent Application 62–146923.
276. Method for producing simple aromatic polythioesterketones, Japan Patent Application 62–119230.
277. Production of aromatic polythioesterketones, Japan Patent Application 62–241922.
278. Method for producing polyarylenesterketones, US Patent 4698393.
279. Method for producing aromatic polymers, US Patent 4721771.
280. Method for producing aromatic simple poly (thio) esterketones, Japan Patent Application 63–317.
281. Method for producing aromatic simple poly (thio) esterketones, Japan Patent Application 63–316.
282. Method for producing polyarylesterketones, US Patent 471611.
283. Gileva, N. G., Solotuchin, M. G., & Salaskin, S. N. (1988). Synthese von aromatis chen Poly ketonen durch Fallungspolukondensation, *Acta Polym. Bild., 39(8)*, 452–455.
284. Lee, I., & Marvel, S. (1983). Polyaromatic esterketones from o, o-disubstituted diphenyl esters, *J. Polym. Sci. Polym. Chem., Ed., 21(8)*, 2189–2195.
285. Method for producing poly arylen ester ketones by means of electrophylic polycondensation, Germany Patent Application 3906178.
286. Colgupoum, H. M., & Lewic, D. F. (1988). Aromatic polyesterketones via superacid catalysis, in "Spec. Polym. 88" Abstracts of the 3-rd International Conference on New Polymeric Materials. Guildford, p. 39.
287. Colgupoum, H. M., Lewic, D. F. (1988). Synthesis of aromatic polyester-ketones in triflouromethanesulphonic acid, *Polym., 29(10)*, 1902.
288. Durvasula, V. R., Stuber, F. A., & Bhattacharyee, D. (1988). Synthesis of Polyphenyleneester and thioester ketones, *J Polym. Sci. A., 27(2)*, 661–669.
289. Method for producing high-molecular polyarylenesulfideketone, US Patent 47182122.

290. Ogawa, T., & Marvel, C. S. (1985). Polyaromatic esterketones and ester-ketone-sulfones having various hydrophilic groups, *J. Polym. Sci.: Polym. Chem.* Ed., *23(4)*, 1231–1241.

291. Percec, V., & Nava, H. (1988). Synthesis of aromatic polyesters by Scholl reaction 1, Poly (1, 1-dinaphthyl ester phenyl ketones), *J. Polym. Sci. A., 26(3),* 783–805.

292. Mitsuree, U., & Nasaki, S. (1987). Synthesis of Aromatic Poly (ester ketones), *Macromolecules, 20(11),* 2675–2677.

293. Method for producing simple polyesterketones, Japan Patent Application, 61–247731.

294. Aromatic polyesterketone and its production, Japan Patent Application, 61–143438.

295. Crystalline polymers with aromatic ketone, simple ether and thioether linkages within the main chain and method for producing same, Japan Patent Application, 61–141730.

296. Producing of crystalline aromatic polysulfidesterketones, Japan Patent Application, 62–529.

297. Producing of crystalline aromatic polyketone with simple ether and sulfide linkages, Japan Patent Application, 62–530.

298. Patel, H. G., Patel, R. M., & Patel, S. R. (1987). Polyketothioesters from 4,4-dichloro acetyl diphenyl ester and their characterization, *J. Macromol Sci. A., 24(7),* 835–340.

299. Method for producing aromatic polyester ketones, Japan Patent Application, 62–220530.

300. Method for producing aromatic polyester ketones, Japan Patent Application, 62–91530.

301. Aromatic co poly ketones and method for producing same, Japan Patent Application, 63–10627.

302. Crystalline aromatic polyester ketones and method for producing same, Japan Patent Application, 61–91165.

303. Aromatic polyester thioester ketones and method for producing same, Japan Patent Application, 61–283622.

304. Producing of polyarylenoxides using carbonates of alkali-earth metals, salts of organic acids and, in some cases, salts of copper as catalysts, US Patent 4774314.

305. Method for producing polyarylenesterketones, US Patent 4767837.

306. Heat-resistant polymer and method for producing same, Japan Patent Application, 62–253618.

307. Heat-resistant polymer and method for producing same. Japan Patent Application, 62–253619.

308. Aromatic polyesterketones, US Patent, 4703102.

309. Producing of aromatic polymers, GB Patent, 1569602.

310. New polymers and method for their production, Japan Patent Application, 61–28523.

311. Aromatic simple polyesterketones with blocked end groups and method for producing same, Japan Patent Application, 61–285221.

312. Aromatic simple polyesterketones and method for their production, Japan Patent Application, 61–176627.

313. Method for producing crystalline aromatic simple polyesterketones, Japan Patent Application, 62–7729.

314. Method for producing fuse aromatic polyesters, US Patent, 4742149.

315. Films from aromatic polyesterketones, Germany Patent Application, 3836169.
316. Method for producing polyarylenestersulfones and polyarylenesterketones, Germany Patent Application, 3836582.
317. Method for producing polyarylenesterketones, Germany Patent Application, 3901072.
318. Polyarylesterketones, US Patent, 4687833.
319. Method for producing oligomer aromatic simple ethers, Poland Patent, 117224.
320. Polymers containing aromatic groupings, GB Patent, 1541568.
321. Corfield, G. C., & Wheatley, G. W. (1988). The synthesis and properties of blok copolymers of polyesteresterketone and polydimethylsiloxane, In "Spec. Polym. 88" Abstracts of the 3-rd International Conference on New Polymeric Materials Cambridge, 68.
322. Method for producing polyarylesterketones, Germany Patent Application, 3700808.
323. Block-copolymers containing polyarylesterketones and methods for their production, US Patent 4774296.
324. Poly (arylesterketones) of improved chain, US Patent, 4767838.
325. New block-copolymer polyarylesterketone-polyesters, US Patent, 4668744.
326. Simple polyarylesterketone block-copolymers, US Patent, 4861915.
327. Producing of polyarylenesterketones by means of consecutive oligomerization and polycondensation in separate reaction zones, US Patent, 4843131.
328. Khasbulatova, Z. S., Kharaev, A. M., Mikitaev, A. K. et al. (1990). *Plast Massy, 11*, 14–17.
329. Khasbulatova, Z. S. (1989). Diversity of methods for synthesizing polyesterketones, in Abstracts of the II Regional Conference, "Chemists of the Northern Caucasus to National Economy" Grozny, 267 [in Russian]
330. Reimer Wolfgang (1999). Polyarylesterketone (PAEK), *Kunststoffe. Bild., 89(10)*, 150, 152, 154.
331. Takeuchi Hasashi, Kakimoto Masa-Aki, & Imai Yoshio (2002). Novel method for synthesizing aromatic polyketones from bis (arylsilanes) and chlorides of aromatic bicarbonic acids, *J. Polym. Sci. A., 40(16)*, 2729–2735.
332. Process for producing polyketones, US Patent 6538098. International Patent Catalogue C 08 П 6/00, 2003.
333. Maeyata Katsuya, Tagata Yoshimasa, Nishimori Hiroki, Yamazaki Megumi, Maruyama Satoshi, & Yonezawa Noriyuki. (2004). Producing of aromatic polyketones on the basis of 2,2{}-diaryloxybisphenyls and derivatives of arylenecarbonic acids accompanied with polymerization with Friedel-Krafts acylation, *React. Funct Polym., 61(1)*, 71–79.
334. Daniels, J. A., Stephenson, J. R. (1995).Producing of aromatic polyketones. GB Patent Application 2287031. International Patent Catalogue C 08 G 67/00.
335. Gibeon Harry W., Pandya Ashish. (1994). Method for producing aromatic polyketones. US Patent 5344914 International Patent Catalogus C 08 G 69/10.
336. Zolotukhin, M. G., Baltacalleja, F. J., Rueda, D. R., & Palacios, J. M. (1997). Aromatic polymers produced by precipitate condensation, *Acta Polym., 48(7)*, 269–273.
337. Zhang Shanjy, Zheng Yubin, Ke Yangchuan, Wu Zhongwen. (1996). Synthesis of aromatic polyesterketones by means of low-temperature polycondensation, *Acta Sci. Nature. Univ. Jibimensis, 1*, 85–88.

338. Hachya Hiroshi, Fukawa Isaburo, Tanabe Tuneaki, Hematsu Nobuyuki, & Takeda Kunihiko. (1999). Chemical structure and physical properties of simple polyesterketone produced from 4,4'-dichlorbenzophenone and sodium carbonate, *Trans. Jap. Soc. Mech. Eng. A., 65(632)*, 71–77.

339. Yang Jinlian, Gibson Harry, W. (1997). Synthesis of polyketones involving nucleophylic replacement through carb-anions obtained from bis (α-aminonitriles), *Macromolecules, 30(19)*, 64–73.

340. Yang Jinlian, Tyberg Christy, S., & Gibson Harry, W. (1999). Synthesis of polyketone containing nucleophylic replacers through carb-anions obtained from bis (α-aminonitriles) Aromatic polyesterketones, *Macromolecules, 32(25)*, 8259–8268.

341. Yonezawa Noriyuki, Ikezaki Tomohide, Nakamura Niroyuki & Maeyama Katsuya (2000). Successful synthesis of all-aromatic polyketons by means of polymerization with aromatic combination in the presence of nickel. Macromolecules, *33(22)*, 8125–8129.

342. Aromatic polyesterketones (2005). US Patent 6909015 International Patent Catalogue C 07 C 65/00.

343. Toriida Masahiro, Kuroki Takashi, Abe Takaharu, Hasegawa Akira, Takamatsu Kuniyuki, Taniguchi Yoshiteru, Hara Isao, Fujiyoshi Setsuko, Nobori Tadahito, & Tamai Shoji (2004). Patent Applicaiton 1464662 International Patent Catalogue C 08 G 65/40.

344. Richter Alexander, Schiemann Vera, Gunzel Berna, Jilg Boris, Uhlich Wilfried (2007). Verfahren zur Herstellung von Polyarylenesterketon, Germany Patent Application, 102006022442 International Patent Catalogue C 08 G 65/40.

345. Chen Liang, Yu Youhai, Mao Huaping, Lu XiaofEng., Yao Lei, & Zhang Wanjin (2005). Synthesis of a new electroactive poly (aryl ester ketone), *Polymer, 46(8)*, 2825–2829.

346. Maikhailin Yu., A. (2007). *Polymer Mater. Articles, Equip. Technol, 5*, 6–15.

347. Sheng Shouri, Kang Yigiang, Huang Zhenzhong, Chen Guohua, Song Caisheng (2004). Synthesis of soluble poly chlor replaced poly aryl ester ketones, *Acta Polym. Sin, 5*, 773–775.

348. Kharaev, A. M., Mikitaev, A. K., & Bazheva, R. Ch. (2007). Halogen-containing polyarylenesterketones, in Proceedings of the 3-rd Russina scientific and practical conference, "Novel polymeric composite materials" Nalchik, 187–190 [in Russian]

349. Liu Baijun, Hu Wei, Chen Chunhai, Jiang Zhenhua, Zhang Wanjin, Wu Zhongwen, Matsumoto Toshihik (2004). Soluble aromatic poly (ester ketones) with a pendant 3, 5-ditrifluoromethylphenyl group, *Polymer, 45(10)*, 3241–3247.

350. Gileva, N. G., Zolotukhin, N. G., Sedova, E. A., Kraikin, V. A., & Salazkin, S. N. (2000). Synthesis of poly arylene phthalidesterketones, in Abstracts of the 2-nd Russian Kargin Symposium "Chemistry and physics of polymers in the beginning of the 21 century" Chernogolovka, Part 1 P. 1/83 [in Russian]

351. Wang Dekun, Wei PEng., & Wu Zhe (2000). Synthesis of soluble polyketones and polyarylenevinylens new reaction of polymerization, *Macromolecules, 33(18)*, 6896–6898.

352. Wang Zhonggang, Chen Tianlu, & Xu Jiping (1995). Synthesis and characteristics of card polyarylesterketones with various alkyl replacers, Acta Polym. Sin, *4*, 494–498.

353. Salazkin, S. N., Donetsky, K. I., Gorshkov, G. V., Shaposhnikova, V. V., Genin Ya.V., & Genina, M. M. (1997). *Vysokomol Soed, A-B, 39,* 1431–1437.

354. Salazkin, S. N., Donetsky, K. I., Gorshkov, G. V., & Shaposhnikova, V. V. (1996). Doklady RAN, *348(1),* 66–68.

355. Donetsky, K. I. (2000). Abstracts of the Thesis for the scientific degree of candidate of chemical sciences. Moscow, 24 p [in Russian]

356. Khalaf Ali A., Aly Kamal L., & Mohammed Ismail (2002). A. New method for synthesizing polymers, *J. Macromol. Sci. A., 39(4),* 333–350.

357. Khalaf Ali A., Alkskas, I. A. (2003). Method for synthesizing polymers, *Eur Polym, J., 39(6),* 1273–1279.

358. Aly Kamal, L. (2004). Synthesis of polymers, *J. Appl. Polym. Sci, 94(4),* 1440–1448.

359. Chu, F. K., Hawker, C. J. (1993). Different syntheses of isomeric hyper branched polyester ketones Polym. Bull, *30(3),* 265–272.

360. Yonezawa Noriyuki, Nakamura Hiroyuki, & Maeyama Katsuya. (2002). Synthesis of all-aromatic poly ketones having controllable isomeric composition and containing links of 2-trifluorometylbisphenylene and 2, 2{}-dimetoxybisphenylene React and Funct. Polym, *52(1),* 19–30.

361. Zhang Shaoyin, Jian Xigao, Xiao Shude, Wang Huiming, & Zhang Jie. (2002). Synthesis and properties of polyarylketone containing bisphthalasinone and methylene groupings, *Acta Polym. Sin., 6,* 842–845.

362. Chen Lianzhou, Jian Xigao, Gao Xia, & Zhang Shouhai. (1999). Synthesis and properties of polyesterketones containing links of chlorphenylphthalasion, *Chin. J. Appl. Chem, 16(3),* 106–108.

363. Gao Ye, & Jian Xi-gao. (2001). Synthesis and crharacterization of polyearyesterketones contnaining 1, 4-naphthaline linkages, *J. Dalian Univ. Technol, 41(1),* 56–58.

364. Wang Mingjing, Liu ChEng., Liu Zhiyong, Dong Liming, & Jian Xigao (2007). Synthesis and properties of polyarylnithilesterketoneketones containing phthalasinon Acta Polym. Sin, 9, 833–837.

365. Zhang Yun-He, Wang Dong, Niu Ya-Ming, Wang Gui-Bin, & Jiang Zhen-Hua (2005). Synthesis and properties of fluor-containing polyarylesterketones with links of 1, 4-naphthylene, Chem. J. Chin Univ., *26(7),* 1378–1380.

366. Kim Woo-Sik, Kim Sang-Youl (1997). Synthesis and properties of polyesters containing naphthalene tetracarboxylic imide, *Macromol Symp., 118,* 99–102.

367. Cao Hui, Ben TEng., Wang Xing, Liu Na, LiuXin-Cai, Zhao Xiao-Gang, Zhang Wan-Jin, & Wei Yen. (2004). Synthesis and properties of chiral polyarylester ketones containing links of 1, 1{}-bis-2-naphtyl, *Chem J. Chin Univ., 25(10),* 1972–1974.

368. Wang FEng., Chen Tianlu, Xu Jiping, Lui Tianxi, Jiang Hongyan, Qi Yinhua, Liu Shengzhou, & Li Xinyu. (2006). Synthesis and characterization of poly(arylene ester ketone) (co)polymers containing sulfonate groups, *Polymer. 47(11),* 4148–4153.

369. Cheng Cai-Xia, Liu-Ling, & Song Cai-Sheng. (2002). Synthesis and properties of aromatic polyesterketoneketone containing carboxylic group within the lateral chain J. Jiangxi Norm Univ Natur Sci. Ed. *26(1),* 60–63.

370. 2, 3, 4, 5, 6-Pentafluorobenzoylbisphenylene ethers and fluor-containing polymers of arylesterketones (2001). US Patent 6172181 International Patetn Catalogue C 08 Π 73/24.

371. Ash, C. E. (1995). Process for producing stabilized polyketones, US Patent 5432220, International Patent Catalogue, C 08 F 6/00.
372. Jiang, Zhen-yu., Huang Hai-Rong, & Chen Jian-Ding (2007).*Synthesis and properties of polyaryl ester ketone and polyarylestersulfone containing link of hexafluoroizopropylydene J. E.* China Univ. Sci. and Technol. Nat. Sci. Ed. *33(3),* 345–349.
373. Yongshen, Xu., Weiguo, Gao., Hongbing, Li, & Guo Jintang (2005).*Synthesis and properties of aromatic polyketones based on CO and stirol or n-ethylstirol J.* Chem. Ind. and Eng. (China). *56(5),* 861–864.
374. Rao, V. L., Sabeena, P. U., Saxena Akanksha, Gopalakrishnan, C., Krishnan, K., Ravindran, P. V., & Ninan, K. N. (2004). *Synthesis and properties of poly(aryl ester ester ketone) copolymers with pendant methyl groups Eur Polym. J. 40(11),* 2645–2651.
375. Tong Yong-Fen, Song Cai-ShEng., Chen Lie, Wen Hong-Li, & Liu Xiao-Ling. (2004). Synthesis and properties of methyl-replaced Polyarylesterketone Chin, *J. Appl.Chem., 21(10),* 993–996.
376. Koumykov, R. M., Vologirov, A. K., Ittiev, A. B., Rusanov, A. L., (2005). Simple aromatic polyesters and polyesterketones based on dinitro-derivatives of clroral in *"Novel polymeric composite materials"*: Proceedings of the 2nd Russian Research-Practical Conference. Nalchik, 225–228 [in Russian].
377. Koumykov, R. M., Bulycheva, E. G., Ittiev, A. B., Mikitaev, A. K., & Rusanov, A. L. (2008). *Plast Massy 3,* 22–24.
378. Polyester ketone and method of producing the same US Patent 7217780 (2006). International Patent Catalogue C 08 G 14/04.
379. Jianying, Li., Yikai, Yu., Mingzhong, Cai., & Song Caisheng (2006). Synthesis and properties of simple polyester ketonesterketone containing lateral Cyano groups Petrochem, *Technol, 35(12),* 1179–1183.
380. Liu, Dan., & Wang, Zhonggang (2008). Novel polyaryletherketones bearing pendant carboxyl groups and their rare earth complexes Part I, *Synthesis and characterization Polymer, 49(23),* 4960–4967.
381. Jeon, In. Yup., Tan Loon-Seng & Baek Jong-Beom (2007).*Synthesis of linear and hyper branched poly(esterketone)s containing flexible oxyethylene spacers Polym. Sci. A. 45(22),* 5112–5122.
382. Maeyama Katsuya, Sekimura Satoshi, Takano Masaomi, & Yonezawa Noriyuki (2004). *Synthesis of copolymers of aromatic polyketones React and Funct Polym, 58(2),* 111–115.
383. Wei, Li., Cai Ming-Zhong, & Song Cai-Sheng (2002). Synthesis of ternary copolymers from 4, 4′-bisphenoxybisphenylsulfone, 4, 4′–bisphenoxybenzophenone and terephthaloyl chloride Chin, *J. Appl. Chem, 19(7),* 653–656.
384. Gao Yan, Dai Ying, Jian Xigao, Peng Shiming, Xue Junmin, & Liu Shengjun (2000). Synthesis and characterization of copolyesterketones produced from hexaphenylreplaced bisphenylbisphenol and hydroquinone. *Acta Polym. Sin, 3,* 271–274.
385. Sharapov, D. S. (2006). *Abstracts of the thesis for the scientific degree of candidate of chemical science* Moscow, 25 p. [in Russian].
386. Kharaeva, R. A., & Ashibokova, O. R. (2005). Synthesis and some properties of Copolyesterketone in Proceedings of Young Scientists Nalchik KBSU, 138–141 [in Russian].

387. Method for preparing polyester copolymers with polycarbonates and polyarylates (2004). US Patent 6815483 International Patent Catalogue C 08 L 67/00.

388. Liu Xiao-Ling, Xu Hai-Yun, & Cai Ming-Zhong (2001). *Synthesis and properties of statistical copolymers of polyesterketoneketone and polyester ketone ester ketone ketone containing naphthalene cycle with on the main chain J.* Jiangxi Norm. Univ Natur Sci. Ed. *25(4),* 292–294.

389. Synthesis and properties of poly (aryl ester ketone) copolymers containing 1, 4-naphthalene moieties J. (2004). *Macromol Sci. A., 41(10),* 1095–1103.

390. Yu Yikai, Xiao Fen, & Cai Mingzhong (2007). Synthesis and properties of poly (arylesterketone ketone)/poly(aryl ester ester ketone ketone) co polymers with pendant cyano groups, *J. Appl. Polym. Sci. 104(6),* 3601–3606.

391. Tong Yong-fen, Song Cai-shEng., Chen Lie, Wen Hong-li, & Liu Xiao-ling. (2005). Synthesis and properties of co polymers of polyarylesterketone containing lateral methyl groupings, *Polym. Mater. Sci. Technol. Eng., 21(4),* 70–72, 76.

392. Gao Yan, Robertson Gilles P, Guiver Michael, D., Mikhailenko Serguei, D., Li Xiang, & Kaliaguine Serge (2004). *Synthesis of copolymers of polyarylene ester ester ketone ketones containing links of naphthalene sulfonic acid within the lateral links, and their use at manufacturing proton-exchange membranes Macromolecules. 37(18),* 6748–6754.

393. Mohwald, Helmut, Fischer Andreas, Frambach Klaus, & Hennig, (2004). In golf that Seven Verfahren zur Herstellung eines zum Protonenaustausch befahigter Polymersystems auf der Basis von Polyarylesterketonen Germany Patent Application 10309135. International Patent Catalogue C 08 G 8/28.

394. Shaposhnikova, V. V., Sharapov, D. S., Kaibova, I. A., Gorlov, V. V., Salazkin, S. N., Dubrovina, L. V., Bragina, T. P., Kazantseva, V. V., Bychko, K. A., Askadsky, A. A., Tkachenko, A. S., Nikiforova, G. G., Petrovskii, P. V., & Peregudov, A. S. (2007). *Vysokomol. Soed 49(10),* 1757–1765.

395. Shaposhnikova, V. V., Salazkin, S. N., Matedova, I. A., & Petrovskii, P. V. Polyarylenether ketones Investigation of approaches to synthesis of amorphous block polymers in abstracts of the 4th International Symposium "*Molecular Order and Mobility in Polymer Systems*" St. Petersburg, 121.

396. Bedanokov, A. Yu. (1999). Abstracts of the thesis for the scientific degree of candidate of chemical sciences Nalchik, 19p [in Russian].

397. Yang Yan-Hua, Dai Xiao-Hui, Zhou Bing, Ma Rong-Tang, & Jiang Zhen-Huang (2005). *Synthesis and characterization of block copolymers containing poly (aryl ester ketone) and liquid crystalline polyester segments Chem. J. Chin. Univ. 26(3),* 589–591.

398. Zhang Yun-He, Liu Qin-Hua, Niu Ya-Ming, Zhang Shu-Ling, Wang Dong, & Jiang Zhen-Hua (2005). Properties and crystallization kinetics of poly (ester ester ketone)-copoly (ester ester ketone ketone) block copolymers, *J. Appl. Polym. Sci. 97(4),* 1652–1658.

399. Polyarylenester ketone phosphine oxide compositions incorporation cycloaliphatic units for use as polymeric binders in thermal control coatings and method for synthesizing same, US Patent 7208551 International Patent Catalogue C 08 L 45/00–216.

400. Keshtov, M. L., Rusanov, A. L., Keshtova, S. V., Pterovskii, P. V., & Sarkisyan, G. B. (2001). Vysokomol Soed. A. *43(12),* 2059–2070.

401. Keshtov, M. L., Rusanov, A. L., Keshtova, S. V., Schegolihin, A. N., & Petrovskii, P. V. (2001). *Vysokomol Soed A, 43(12)*, 2071–2080.
402. Brandukova Natalya, E., & Vygodskii Yakov, S. (1995). Novel Poly α-diketones and co polymers on their base, J. *Macromol Sci. A, 32*, 941–950.
403. Yandrasits, M. A., Zhang, A. Q., Bruno, K., Yoon, Y., Sridhar, K., Chuang, Y. W., Harris, F. W., & ChEng., S. Z. D. (1994). *Liquid-crystal polyenamineketones produced via hydrogen bonds. Polym. Int., 33(1)*, 71–77.
404. Yongli, Mi., Zheng Sixun, Chan Chi-ming, & Guo Qipeng (1998). Mixes of phenolphthalein with thermotropic liquid crystal co polyester, *J. Appl. Polym. Sci., 69(10)*, 1923–1931.
405. Arjunan Palanisamy (1995). Productions of Polyesters from Polyketones, US Patent 5466780. International Patent Catalogue C 08 F 8/06, C 08 K 5/06.
406. Arjunan Palanisamy (1996). Process of transformations of poly ketones into the complex polyesters US Patent, 55506312. International Patent Catalogue C 08 F 20/00.
407. Matyushov, V. F., & Golovan, S. V. (2003). Method for producing nonsaturated oligoarylesterketones, RF Patent 2201942. International Patent Catalogue C 08 G 61/12.
408. Matyushov, V. F., & Golovan S. V. (2003). Method for producing nonsaturated oligoarylester ketones, RF Patent Application 2001109440/04 International Patent Catalogue C 08 G 61/12.
409. Matyushov, V. F., Golovan, S. V., & Malisheva, T. L. (2000). Method for producing oligoarylester ketones with end amino-groups Ukraine Patent 28015. International Patent Catalogue C 08 G 8/02.
410. Zhaobin Qiu, Zhishen Mo, & Hongfang Zhang (2000). *Synthesis and crystalline structure of oligomer of arylesterketone Chem Res, 11*, 5–7.
411. Guo Qingzhong, & Chen Tianlu (2004). Synthesis of macrocyclic oligomers of aryleneketones containing phthaloyl links by means of Friedel Crafts acylation reaction, *Chem. Lett., 33(4)*, 414–415.
412. Wang Hong Hua, Ding Jin, & Chen Tian Lu. (2004). Cyclic oligomers of phenolphthale in polyarylene ester sulfone (ketone): preparation through cyclo-depolymerization of corresponding *polymers Chin. Chem. Lett, 15(11)*, 1377–1379.
413. Kharaev, A. M., Basheva, R. Ch., Istepanova, O. L., Istepanov, M. I., & Kharaeva, R. A. (2000). Aromatic oligoesterketones for polycondensation, RF Patent, 2327680 International Paten Catalogue C 07 C 43/02–2006.
414. Bedanokov, A. Yu., Shaov, A. Kh., Kharaev, A. M., & Dorofeev, V. T., *Plast Massy 4*, 42.
415. Bedanokov Azamat, U., Shaov Abubekir Ch., Charaev Arsen, M., & Mashukov Nurali, I. (1997). Synthesis and some properties of oligo-and polyesterketones based on bisphenylpropane in Proceedings of International Symposium "New Approaches in Polymeric Syntheses and Macromolecular Formation." Sankt-Petersburg, 13–17 [in Russian].
416. Kharaev, A. M., Bazheva, R. C., h., Kazancheva, F. K., Kharaeva, R. A., Bahov, R. T., Sablirova, E. R., & Chaika, A. A. (2005). Aromatic polyester ketones and polyesteresterketones as perspective thermostable constructional materials, in Proceedings of the 2-nd Russian Research-Practical Conference Nalchik, 68–72 [in Russian].

417. Kharaev, A. M., Bazheva, R. C., h., Kharaeva, R. A., Beslaneeva, Z. L., Pampuha, E. V., & Barokova, E. B. (2005). Producing of polyesterketones and polyesteresterketones on the basis of bisphenols of various composition in Proceedings of the 2-nd Russian Research-Practical Conference. Nalchik, 44–47 [in Russian].

418. Bazheva, R. C., h., Kharaev, A. M., Olhovaia, G. G., Barokova, E. B., & Chaika, A. A. (2006). Polyester-polyesterketone block-copolymers in Abstracts of the international Conference on Organic Chemistry "Organic Chemistry from Butlerov and Belshtein till nowadays" Sankt-Petersburg, 716 [in Russian].

419. Aromatic polymers GB Patent 1563223.

420. Polysulfoneesterketones Germany Patent Application 3742445.

421. Germany Patent Application 3742264.

422. Aromatic polymers Macromolecules (1984). *17(1),* 10–14.

423. Khasbulatova, Z. S., Kharaev, A. M., & Mikitaev, A. K. (2009). Khim Prom Segodnya, 10, 29–31.

424. Wen Hong Li, Song Cai-ShEng, Tong Yong-Fen, Chen Lie, & Liu Xiao-Ling. (2005). Synthesis and properties of poly(aryl ester sulfone ester ketone ketone) (PESEKK) *J. Appl. Polym. Sci. 96(2),* P. 489–493.

425. Li Wei, Cai Ming-Zhong Ying Chin. (2004). *J. Appl. Chem. 21(7),* 669–672.

426. Sheng Shou-Ri, Luo Qiu-Yan, Yi-Huo, Luo Zhuo, & Liu Xiao-Ling (2008). Song Cai-Sheng Synthesis and properties of novel organosoluble aromatic poly (ester ketone)s containing pendant methyl groups and sulfone linkages, *J. Appl. Polym. Sci, 107(1),* 683–687.

427. Tong Yong-fen, Song Cai-shEng., Wen Hong-li, Chen Lie, & Liu Xiao-Ling (2005). Synthesis and properties of copolymers containing methyl replacers *Polym. Mater. Sci. Technol, 21(2),* 162–165.

428. Sheng Shou-ri, Luo-Qiu-yan, Huo Yi, Liu Xiao-ling, & Pei Xue-liang (2006). Song Cai-sheng Synthesis and properties of soluble methyl-replaced polyarylesterketonestersulfonesterketones, *Polym. Mater. Sci. Technol., 22(3),* 85–87, 92.

429. Xie Guang-Liang, Liao Gui-Hong, & Wu Fang-Juan (2008). Song Cai-Sheng Synthesis and adsorption properties of poly (arylestersulfonesterketone)ketone with lateral carboxylic groups *Chin. J. Appl. Chem., 25(3),* 295–299.

430. Charaev, A. M., Khasbulatova, Z. S., Basheva, R. Ch., Kharaeva, R. A., Begieva, M. B., Istepanova, O. L., & Istepanov, M. I. (2007). Izv Vuzov Sev Kav Reg Estestv. *Nauki 3,* 50–52.

431. Chen Lie, Song Cai-ShEng., Wen Hong-Li, Tong Yong-Fen, & Liu Xiao-Ling (2004). Synthesis of statistical polyester sulfonesterketoneketones containing bis (o-methyl) groups Chin., J. *Appl. Chem. 21(12),* 1245–1248.

432. Arthanareeswaran, G., Mohan, D., & Raajenthiren, M. (2007). Preparation and performance of polysulfone-sulfonated poly (ester ester ketone) blend ultrafiltration membranes. Part I *Appl. Surface Sci. 253(21),* 8705–8712.

433. Xing Peixiang, Robertson Gilles, P., Guiver Michael, D., Mikhailenko Serguei, D., & Kaliaguine Serge (2004). Sulfonated poly (aryl ester ketones) containing naphthalene moieties for proton exchange membranes J. *Polym. Sci. A. 42(12),* 2866–2876.

434. Khasbulatova, Z. S., & Shustov, G. B. (2009). Aromatic oligomers for synthesing polyesters in proceedings of the X International Conference of chemistry and physic-chemistry of oligomers, Volgograd, 99 [in Russian].
435. Khasbulatova, Z. S., Shustov, G. B., & Mikitaeva, A. K. (2010). Vysokomol. Soed B *52(4)*, 702–705.

CHAPTER 8

CARBON NANOFIBERS FOR ENVIRONMENTAL REMEDIATION— A COMPREHENSIVE REVIEW

SAEEDEH RAFIEI, BABAK NOROOZI, and A. K. HAGHI

CONTENTS

ABSTRACT

Carbon nanofibers (sometimes known as carbon nanofilaments or CNF) can be produced in a relative large scale by electrospinning of Polyacrylonitrile solution in dimethyl formaldehyde, stabilization and carbonization process. The porosity, pore volume and surface area of carbon nanofiber enhanced during the chemical or physical activation. This paper is a review of electrospinning method of CNF production and heat pretreatments and activations process to make activated carbon nanofiber. Attention is also given to some of the possible applications of this nanostructures which center around the unique blend of properties exhibited by the material, which include: hydrogen adsorption properties, energy storage media, catalyst support, and regenerative medicine.

8.1   INTRODUCTION

In recent years, porous materials have been of immense interest because of their potential for applications in various fields, ranging from chemistry to physics, and to biotechnology [1]. Introducing nanometer-sized porosity has been shown as an effective strategy to achieve desired properties while maintaining micro structural feature sizes commensurate with the decreasing length-scales of devices and membranes [2]. On the other hand, fibrous materials have filled many needs in many areas due to their intrinsically high surfaces, interfiber pores and engineering versatility. In concept, coupling nano porosity in the ultra-fine nano fibers should lead to the highest possible specific surface and fibrous materials. Hence, the synthesis of organic polymer or inorganic nano fibers with nano porous structures might be very useful in a widely areas, such as membranes technology [3], tissue engineering, drug delivery [4], adsorption materials [5–8], filtration and separation [9], sensors [10], catalyst supports [11, 12] and electrode materials [13–17] and so forth.

Activated carbon fibers (ACF) and nano fibers (ACNF) are a relatively modern form of porous carbon material with a number of significant advantages over the more traditional powder or granular forms. Advantages include high adsorption and desorption rates, thanks to the smaller fiber diameter and hence very low diffusion limitations, great adsorption capacities at low concentrations of adsorbates, and excellent flexibility [18, 19].

The adsorption and the isolation technique is an efficient method in environmental problems, the toxic materials in sorts of industrial wastewater and exhaust fumes can be absorbed by using various adsorbents, so that the fumes and the liquid are up to standard of environmental protection. The key problem of the adsorption and the isolation technique lies in the adsorbents; the commonly used adsorbents are activated carbon, silica gel, acid terra alba and zeolite molecular sieve, etc. [20–22]. But, not only the adsorption characterization of these materials but also the operating characterization and the reproducing ability of these materials are all very weak. So searching for a high quality adsorption material has become a subject concerned by experts all over the world.

Although several methods have been proposed for nano fiber manufacturing so far, an efficient and cost effective procedure of production is still a challenge and is debated by many experts [23]. Some of different methods for nano fiber manufacturing are Drawing [24], Template synthesis [25], Phase separation [26], Self-assembly [27] vapor growths, arc discharge, laser ablation and chemical vapor deposition [28]. One of the most versatile methods in recent decades is electro spinning which is a broadly used technology for electrostatic fiber formation which uses electrical forces to produce polymer fibers with diameters ranging from several nm to several micrometers using polymer solutions of both natural and synthetic polymers has seen a tremendous increase in research and commercial attention over the past decade [29–31]. Electro spinning is a remarkably simple and versatile technique to prepare polymers or composite materials nanofibers. As many research showed that, by changing appropriate electro spinning parameters, it can be possible to obtain the fibers with nano porous structures, for which the basic principle based on away that phase separation processes take place during electrospinning [1, 30, 32]. Nanomaterials are traditionally defined to be those with at least one of the three dimensions equal to or less than 100 nm, 8 where as an upper limit of 1000 nm is usually adopted to define nano fibers. A large number of special properties of polymer nano fibers have been reported due to the high specific surface area and surface area to volume ratio. In contrast to the normal PAN fibers with diameters of about 20μm [33–34], electro spun PAN nano fibers with diameters in the submicron range result in activated carbon nano fibers (ACNF) with very much higher specific surface area [1, 35].

ACNF is an excellent adsorbent and has found usage in applications such as gas-phase and liquid phase adsorption [5, 20, 36] as well as electrodes for super capacitors and batteries [14, 37]. The literature review concerning ACNF is summarized as follows:

Kim and Yang [14] prepared ACNF from electro spun PAN nano fibers activated in steam at 700–800 °C and found that the specific surface area of the ACNF activated at 700 °C was the highest but the Mesopore volume fraction was the lowest. However, the work by Lee et al. [38] showed an opposite result. Song et al. [21] investigated the effect of activation time on the formation of ACNF (ultra-thin PAN fiber based) activated in steam at 1000 °C. Ji et al. [39] made mesoporous ACNF produced from electrospun PAN nano fibers through physical activation with silica and conducted chemical activations by potassium hydroxide and zinc chloride to increase specific surface area and pore volume of ACNF [40]. It must be pointed out that different methods and conditions of activation lead to very different physical properties and adsorption capacities for ACNF.

This review provides an overview of the most preferable production methods of carbon fiber and nanofiber and activated form of them. Because of the extraordinary combination of physical and chemical properties exhibited by carbon nanofibers and activated form of it, which blends two properties that rarely coexist: high surface area and high electrical conductivity, which are the result of the unique stacking and crystalline order present within the structure, there are tremendous opportunities to exploit the potential of this form of carbon in a number of areas, some of which are discussed in this paper.

8.2 CARBON FIBERS

The existence of carbon fiber (CFs) came into being in 1879 when Thomas Edison recorded the use of carbon fiber as a filament element in electric lamp. Fibers were first prepared from rayon fibers by the US Union Carbide Corporation and the US Air Force Materials Laboratory in 1959 [41]. In 1960, it was realized that carbon fiber is very useful as reinforcement material in many applications. Since then a great deal of improvement has been made in the process and product through research work carried out in USA, Japan and UK. In 1960s, High strength Polyacrylonitrile (PAN) based carbon fiber was first produced in Japan and UK and pitch based carbon fiber in Japan and USA.

Carbon fibers can be produced from a wide variety of precursors in the range from natural materials to various thermoplastic and thermosetting precursors Materials, such as Polyacrylonitrile (PAN), mesophase pitch, petroleum, coal pitches, phenolic resins, polyvinylidene chloride (PVDC), rayon (viscose), etc. [42–43]. About 90% of world's total carbon fiber productions are polyacrylonitrile (PAN)-based. To make carbon fibers from PAN precursor, PAN-based fibers are generally subjected to four pyrolysis processes, namely oxidation stabilization, carbonization and graphitization or activation; they will be explained in following sections later [43].

Among all kind of carbon fibers, PAN-based carbon fibers are the preferred reinforcement for structural composites with the result of their excellent specific strength and stiffness combined with their light weight as well as lower cost, butin general, PAN-based carbon fibers have lower carbonization yield than aromatic structure based precursors, such as pitch, phenol, polybenzimidal, polyimide etc. The carbon yield strongly depends on the chemical and morphological structures of the precursor fibers [44].

Carbon fibers are expected to be in the increasing demand for composite materials in automobile, housing, sport, and leisure industries as well as airplane and space applications [45]. They requires high strength and high elasticity, the high strength type is produced from Polyacrylonitrile (PAN) and the high elasticity type is manufactured from coat tar pitch. In order to meet expanded use in some high-tech sectors, many novel approaches, such as dry-wet spinning [46], steam drawing [34], increasing the molecular weight of precursors polymer [47], modifying the precursors prior to stabilization [43–47], etc., have been performed to increase the tensile strength of PAN-based carbon fibers.

The quality of the high performance carbon fibers depends mainly on the composition and quality of the precursor fibers. In order to obtain high performance PAN-based carbon fibers, the combination of both physical mechanical properties and chemical composition should be optimized [48]. Producing a high performance

PAN-based carbon fiber and activated carbon fiber is not an easy task, since it involves many steps that must be carefully controlled and optimized. Such steps are the dope formulation, spinning and post spinning processes as well as the pyrolysis process. At the same time, there are several factors that need to be considered in order to ensure the success of each step. However, among all steps, the pyrolysis process is the most important step and can be regarded as the heart of the carbon fiber production [43].

Carbon fibers possess high mechanical strengths and module, superior stiffness, excellent electrical and thermal conductivities, as well as strong fatigue and corrosion resistance; therefore, they have been widely used for numerous applications particularly for the development of large load-bearing composites. Conventional carbon fibers are prepared from precursors such as Polyacrylonitrile (PAN) [49, 50]. Ch. Kim et al. [44] produced two-phase carbon fibers from electrospinning by pitch and PAN precursor. He proved that the fiber diameter, the Carbon yield and the electrical conductivity are increased with increasing Pitch component.

Carbon fibers have various applications because of their porous structure [51]. The preparation of drinkable, high quality water for the electronics and pharmaceutical industries, treatment of secondary effluent from sewage processing plants, gas separation for industrial application, hemo dialyzers, and the controlled release of drugs to mention only a few applications [49–52].

8.2.1 CLASSIFICATIONS OF CARBON FIBERS

The manufacturing technology for carbon fibers is based on the high-temperature pyrolysis of organic compounds, conducted in an inert atmosphere. Some carbon fiber classifications are based on the magnitude of the final heat-treatment temperature (HTT) during production of the carbon fibers through pyrolysis, and also on the carbon content of the final product. Accordingly, carbon fibers may be subdivided into three classes: partially carbonized fibers (HTT 500 °C, carbon content up to 90 wt%), carbonized fibers (HTT 500–1500 °C, carbon content 91–99 wt%), and graphitized fibers (HTT 2000–3000 °C, carbon content over 99 wt%). Carbon fibers can be classified by raw materials as well, for example, rayon-based fibers, PAN-based fibers, pitch-based mesophase fibers, lignin based fibers and gas phase production fibers [52].

8.2.2 PROPERTIES OF CARBON FIBERS

The characteristics of carbon fiber material are influenced by choices of the initial polymer raw material, conditions of carbonization and heat treatment, and also by introduction of certain additives.

8.2.2.1   MECHANICAL PROPERTIES

PAN-based carbon fibers demonstrate 200–400 GPa Young's modulus upon longitudinal stretching of the fiber, while upon transverse stretching the Young's modulus is 5–25 GPa, and the compressive strength is 6 GPa. It has been suggested that the Young's modulus of carbon fibers depends on the orientation of graphite crystallites in the carbon fiber, while the strength is determined by the intra fiber bonding [52].

8.2.2.2   CHEMICAL STABILITY

An important property of carbon fibers that largely determines their prospective uses in many fields is their stability with respect to aggressive agents. This property is related to some structural features of carbon fibers and primarily depends on the type of initial raw material, the heat-treatment temperature, and the presence of element in the fiber. It is evident that the acid stability of carbon fibers increases with increase in heat treatment temperatures as the proportion of the more stable bonds increases, while the more perfect carbon structure excludes reagent diffusion into the fiber matrix. While at room temperature there has been little change observed in carbon fibers even after prolonged periods of exposure to corrosive liquids. The stability of carbon fibers at elevated temperatures decreases, especially if the reagents are oxidizing (i.e., nitric acid, sodium hypo chloride) [52].

8.2.2.3   APPLICATIONS OF CARBON FIBERS

The expansion of the areas of application for carbon fibers is stimulated by their attractive properties, not found in other materials, such as strength, electrical conductivity, stability on exposure to reactive media, low density, low-to-negative coefficient of thermal expansion, and resistance to shock heating. The most representative applications of carbon fibers and element carbon fibers are as sorption materials, electrostatic discharge materials, catalysts, and reinforcement materials in composites.

8.3 POLYACRYLONITRILE (PAN)

It is well know that PAN is made from acrylonitrile, which was prepared by Moureu in 1893. The chemical structure of PAN is illustrated in Fig. 8.1. It is a resinous, fibrous, or rubbery organic polymer and can be used to make acrylic fibers. PAN is sometimes used to make plastic bottles, and as a starting material for making carbon fibers. It is chemically modified to make the carbon fibers found in plenty of both high-tech and common daily applications such as civil and military aircraft primary and secondary structures, missiles, solid propellant rocket motors, pressure vessels, fishing rods, tennis rackets, badminton rackets & high-tech bicycles. Homopolymer of PAN has been used as fibers in hot gas filtration systems, outdoor awnings, sails for yachts, and even fiber reinforced concrete. The homopolymer was developed for the manufacture of fibers in 1940, after a suitable solvent had been discovered by DuPont in the USA, while Bayer developed an aqueous based solution [52]. Almost all polyacrylonitrile resins are copolymers made from mixtures of monomers; with acrylonitrile as the main component. It is a component repeat unit in several important copolymers, such as styrene-acrylonitrile or SAN and ABS plastic. Copolymers containing polyacrylonitrile are often used as fibers to make knitted clothing, like socks and sweaters, as well as outdoor products like tents and similar items. If the label of a piece of clothing says "acrylic," then it is made out of some copolymer of polyacrylonitrile. It was made into spun fiber at DuPont in 1941 and marketed under the name of Orlon. Acrylonitrile is commonly employed as a comonomer with styrene (e.g., SAN, ABS, and ASA plastics).

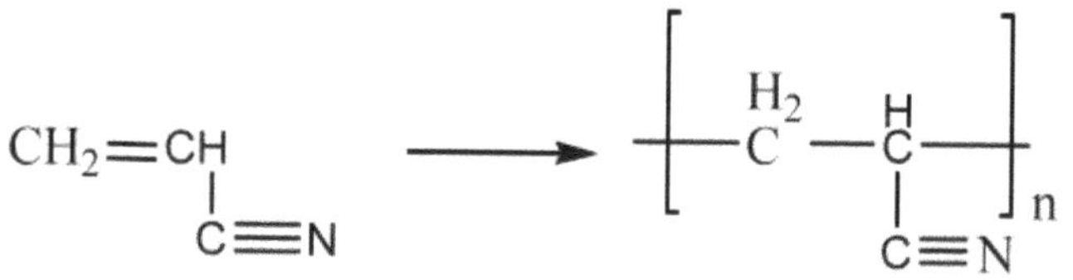

FIGURE 8.1 Chemical structure of PAN.

PAN fibers are used in weaving (blanket, carpet and clothes) and in engineering-housing (instead of asbestos) and most importantly for producing carbon fibers [53]. In recent decade PAN fibers are considered as main material for production of carbon fibers. PAN fibers manufactured presently are composed of at least 85% by weight of acrylonitrile (AN) units. The remaining 15% consists of neutral and/or ionic comonomers, which are added to improve the properties of the fibers. Neutral comonomers like methyl acrylate (MA), vinyl acetate (VA), or methyl methacrylate (MMA) are used to modify the solubility of the PAN copolymers in spinning solvents, to modify the PAN fiber morphology, and to improve the rate of diffusion of dyes into the PAN fiber. Ionic and acidic comonomers including the sulfonate groups like SMS, SAMPS, sodium p-styrene sulfonate (SSS), sodium p- sulfophenyl methallyl ether (SMPE), and IA also can be used to provide dye sites apart from end groups and to increase hydrophilicity [54].

8.4   ACTIVATED CARBON FIBERS

One of the disadvantages of carbon fibers is its low surface area This fact restrict the applications of these materials like hydrogen (or energy) storage or treatment of water and waste water; therefore is necessary increase the surface area to improve the yield in these materials [55]. Activation treatment greatly increases the number of micropores and mesopores [18].

Chemical activation with KOH or NaOH is an effective method to prepare activated carbon materials [55–56]. Chemical agent activation soaks the carbon material using chemical agent, during the process of heating and activating, carbon element will liberate with tiny molecule such as CO or CO_2. $ZnCl_2$, KOH, H3PO4 [57] are the commonly used chemical agents [22].

Chemical activation presents several advantages and disadvantages compared to physical activation. The main advantages are the higher yield, lower temperature of activation, less activation time and generally, higher development of porosity. Among the disadvantages, the activating agents are more expensive (KOH and NaOH vs. CO_2 and H_2O) and it is also necessary an additional washing stage. Because the character of ACF manufactured by this method is unstable, we seldom use it [22]. Moreover, these hydroxides are very corrosive. Physical activation with carbon dioxide or steam is the usual procedure to obtain activated carbon fibers (ACF). Chemical activation of carbon fibers by $ZnCl_2$, $AlCl_3$, H_3PO_4, H3BO3, has been reported [58].

Activated Carbon Fiber is one member of carbon family with multiple holes, which has its unique performance. It has the characters such as big specific surface area, developed microcellular structure, large adsorptive capacity, fast absorbed and desorbed velocity and easy regeneration capacity, etc. they have begun to find use in fluid filtration applications as an alternative to activated carbon granules. Their properties differ from granules with respect to porosity and physical form, which can confer certain advantages, e.g., their smaller dimensions give improved access of the adsorptive and the micropores are accessible from the surface. Adsorption rates of organic vapors are therefore faster than in granules where diffusion through macropores and mesopores must occur first. A further advantage is their ability to be formed into both woven and nonwoven mats, where problems due to channeling when granules are used are avoided. Materials of this type can act as aerosol and particulate filters, as well as microporous adsorbents [59]. According to the aperture classification standard by IUPAC, the pores can be classified into three categories, namely micropore (pore size < 2 nm), mesopore (2 nm < pore size < 50 nm) and macropore (pore size >50 nm). It is proved that mesopores or macropores in ACF can thus improve the adsorption effectivity for larger molecules or macromolecules such as protein and virus [19].

The PAN based activated carbon hollow fibers (ACHF) have brought on many investigators' interest, since PAN-based ACHF shows large adsorption capacity. PAN hollow fibers are pretreated with ammonium dibasic phosphate and then further oxidized in air, carbonized in nitrogen, and activated with carbon dioxide.

One of the important applications of ACF is to remove formaldehyde from atmosphere. It has been investigated for many years. It was found that porous carbons derived from PAN showed larger formaldehyde adsorption capacity, which originated from its abundant nitrogen functionalities on the surface [20].

8.4.1 METHODS OF ACTIVATED CARBON FIBER PREPARATION

Fibers are formed from above mentioned precursors, which are then subject to various heat treatments in controlled atmospheres to yield carbon fibers, often with specific mechanical properties. A final step in this procedure can be either chemical or physical activation of the carbon fibers. In general, nongraphitizable carbons, which are more disordered can be activated physically in steam or CO whereas the more ordered graphitizable carbons require chemical activation for the generation of porosity [60].

As a result activated carbon fiber is produced through a series of process consisting of stabilization, carbonization, and activation of precursor fibers. It is important to improve the efficiency of the production process as well as to select low cost precursors [42]. The stabilization process, air oxidation of precursor fibers at 200–300 °C, is the process required to prevent the precursor fibers from melting during the subsequent carbonization process [61, 62]. It is essential for PAN and pitches, but is not essential for phenolic resin and cellulose, because the latter precursors are thermosetting resins. Phenolic resin is known to produce higher surface area ACF as compared with other precursors. It is, therefore, very advantageous if we could improve the production efficiency of phenol resin based ACF by, possibly, simple and cost effective methods.

8.4.2 PROPERTIES OF ACTIVATED CARBON FIBER (ACF)

Some of the most important properties of ACF are listed as below Table 8.1.

TABLE 8.1   Activated Carbon Fiber Properties

Most important properties of ACF	References
Fast speed of adsorption and desorption as a result of large surface area and average aperture of micropore.	[63]
ACF also has excellent adsorption to low concentration substance, because its adsorption forces and unit adsorptive volume.	[64, 65]
The capability of turning into different patterns, like paper, nonwoven, a honeycomb structure and corrugated cardboard because of good tensile strength.	[63]
ACF itself is not easy to become powder so it will not cause second pollution.	[63]
ACF's thin and light adsorption layer allow it be used in the small treatment with high efficiency.	[66]

TABLE 8.1 *(Continued)*

Most important properties of ACF	References
ACFs adsorption can reach to the expectable efficiency in short time because it's Low density and small loss of press.	[66]
ACF can also be applied as fuel cell electrodes material.	[67]
Without reducing its adsorption function, it's easy to be regenerated. And it can be used for very long time.	[67]
ACF can be a deoxidizer to recycle the precious metal.	[66]
ACFs have acid-proof and alkali-proof properties.	[68, 69]

8.4.3 APPLICATION OF ACF

ACFs have been successfully applied in many fields, such as the treatment of organic and inorganic waste gases, the recovery of organic solvent, air cleaning and deodorization, treatment of wastewater and drinking water, separation and recovery of precious metals, as medical adsorbents and protective articles, and in electrodes [57]. Some of the most common applications of ACF are listed below Table 8.2.

TABLE 8.2 Activated Carbon Fiber Applications

ACF applications	Explanation	Reference
Recovery of Organic Compounds and Solvents	ACF can be used in gas/air separation, recovery of organic compounds and solvents, especially for caustic nitrides, and low boiling point solvent.	[64, 70]
Air purification	ACF can eliminate malodorous substance in the air, especially for aryl substance which will generate carcinogens.	[57]
Wastewater treatment	ACF is suitable in organic waste-water treatment, like substance content phenol, medical waste, etc. which are hard to decompose by organism. With large quantity of adsorption volume, fast speed of adsorption, excellent desorption function, and easy to regenerate. ACF can be used in a small, continuous, simple design condition, which cost low and do not make second pollution.	[9]

TABLE 8.2 *(Continued)*

ACF applications	Explanation	Reference
Water purification	They can be applied in deodorizing and de-coloring Applications in different industrial field, like food and beverage, pharmaceutical, sugar-making, wine-making and also for super-pure water treatment system of electronic industry and aqueous filtration treatment.	[9, 71]
Domestic products	ACF can be a refrigerator deodorizer and keep food fresh	[57]

8.5 ACTIVATED CARBON

Needs for porous materials in industrial application and in our daily life are increasing. Porous carbon materials, especially those containing micropores or mesopores, are being used in various applications such as adsorbent and catalyst supports. Activated carbon has played a major role in adsorption technology over the last few years [72]. Porous carbons have highly developed porosity and an extended surface area. Their preparation involves two methods: a physical method that initiates the activation at high temperature with CO_2 or a water steam stream, and a chemical method that initiates the activation using the micro explosion behaviors of chemical agents [73]. Normally, all carbonaceous materials can be converted into porous carbons, although the properties of the final products will differ according to the nature of the raw materials used.

Toxic organic compounds, such as aromatics and chlorinated hydrocarbons in the gaseous or liquid effluent streams from the production of chemicals and pharmaceuticals are normally removed by adsorption with activated carbons (ACs) in a fixed-bed purifier. However, both contaminant and water molecules are adsorbed at the same time owing to the hydrophilic character acquired during the production process of AC, thus limiting the capacity of the adsorbent which is directly related to its highly micro porous structure, which unfortunately leads to weak mechanical strength resulting in decrepitation into fines in practical application [74]. Most commercially available activated carbons are extremely micro porous and of high surface area, and consequently they have high efficiency for the adsorption or removal of low-molecular weight compounds. The mesoporous activated carbons are expected to be excellent adsorbents for the removal and recovery of mesomolecular weight compounds.

8.6   CARBON NANOSTRUCTURES

8.6.1   CARBON NANOTUBES (CNTS)

Nanotubes of carbon and other materials are arguably the most fascinating materials playing an important role in nanotechnology today. One can think of carbon nano tubes as a sheet of graphite rolled into a tube with bonds at the end of the sheet forming the bonds that close the tube. Due to their small size and their extraordinary physicochemical properties, much attention has been paid to the interesting sp2-based fibrous carbons, including carbon nano tubes. It is generally accepted that carbon nanotubes consist of single or multiple grapheme sheets rolled into concentric cylinders: thus giving rise to single wall carbon nanotubes (SWNTs) or multiwall carbon nanotubes (MWNTs). A single-walled nanotubes (SWNT) can have a diameter of 2 nm and a length of 100pm, making it effectively a one dimensional structure called a nanowires [75].

Their unique mechanical, electronic, and other properties are expected to result in revolutionary new materials and devices. These nanomaterials, produced mostly by synthetic bottom up methods, are discontinuous objects, and this leads to difficulties with their alignment, assembly, and processing into applications. Partly because of this, and despite considerable effort, a viable carbon nanotubes reinforced super nanocomposite is yet to be demonstrated. Advanced continuous fibers produced a revolution in the field of structural materials and composites in the last few decades as a result of their high strength, stiffness, and continuity, which, in turn, meant processing and alignment that were economically feasible [76]. Fiber mechanical properties are known to substantially improve with a decrease in the fiber diameter. Hence, there is a considerable interest in the development of advanced continuous fibers with nanoscale diameters. Electrospinning technology enables production of these continuous polymer nanofibers from polymer solutions or melts in high electric fields [23, 77].

8.6.2 CARBON NANOFIBERS (CNF)

Carbon nanofibers, like other one-dimensional (1D) nanostructures such as nanowires, nanotubes, and molecular wires, are receiving increasing attention because of their large length to diameter ratio. With the development of nanotechnology in fiber fields, carbon nanofibers (CNFs) gradually attracted much attention after the discovery of carbon nanotubes by Iijima in 1991 [78]. CNF cost significantly less to produce than carbon nanotubes (CNT) and therefore offer significant advantages over nanotubes for certain applications, providing a high performance to cost ratio [28, 79].

CNFs are currently widely used in many fields, such as reinforcement materials, gas adsorption/desorption, rechargeable batteries, templates for nanotubes, high-temperature filters, supports for high-temperature catalysis, nanoelectronics and supercapacitors, due to their high aspect ratio, large specific surface area, high-temperatures resistance and good electrical/thermal conductivities. The conventional preparation methods for CNFs, including the substrate method, spraying method, vapor growth method and plasma-enhanced chemical vapor deposition method, are known to be very complicated and costly. Therefore, a simple and inexpensive electrospinning process, first patented by Cooley in 1902, 2 has been increasingly used as the optimum method to fabricate continuous CNFs during the last decade [41]. The primary characteristic that distinguishes CNF from CNT resides in grapheme plane alignment: if the graphene plane and fiber axis do not align, the structure is defined as a CNF, but when parallel, the structure is considered as a CNT [79].

Carbon nanofibers with controllable nanoporous structures can be prepared via different ways. An oldest method is the catalytic decomposition of certain hydrocarbons on small metal particles such as iron, cobalt, nickel, and some of their alloys [80]. The mechanism includes hydrocarbon adsorption on a metal surface, conversion of the adsorbed hydrocarbon to adsorbed surface carbon via surface reactions, subsequent segregation of surface carbon into the layers near the surface, diffusion of carbon through metal particles, and then precipitation on the rare side of the particle [81]. The size of the catalyst nanoparticles seems to be the determining factor for the diameter of the carbon nanostructures grown on it. Small nanoparticles catalyze this grown better than the big ones due to that exhibit peculiar electronic properties (and thus catalytic properties) as consequence of their unusual ratio surface atom/bulk atom [80, 82, 83].

In a typical operation, about 100 mg of the powdered catalyst is placed in a ceramic boat, which is positioned in a quartz tube located in a horizontal tube furnace. The sample is initially reduced in a 10% hydrogen helium stream at 600 °C and then quickly brought to the desired reaction temperature. Following this step, a predetermined mixture of hydrocarbon, hydrogen, and inert gas is introduced into the system and the reaction allowed proceeding for periods of about 2 h [6].

Carbon nanofibers were successfully prepared via electrospinning of PAN solution [78, 84, 85]. Electrospinning is a unique method for producing nanofibers or ultrafine fibers, and uses an electromagnetic field with a voltage sufficient to overcome the surface tension. Electrospun fibers have the characteristics of high specific surface area, high aspect ratio, dimensional stability, etc. [20].

V. Barranco.et al [86] divided carbon nanofiber into two groups, (i) highly graphitic CNFs; and (ii) lowly graphitic ones. He investigated that the lowly graphitic CNFs have been obtained by certain procedures:
- from blends of polymers in which a polymer acts as the carbon precursor and the other gives way to porosity, the latter being removed along the carbonization process;
- by electrospinning of precursors, this gives way to webs of CNFs after carbonization;
- by using an anodic alumina as a template, which leads to CNFs with marked mesoporosity;
  - from electrochemical decomposition of chloroform; and
  - from flames of ethanol.

The two last procedures lead to nanofibers with high oxygen contents. In all cases, the low crystallinity of the CNFs could be an advantage for subsequent activation, and thus, activated CNFs with a larger specific surface area and a higher specific capacitance can be obtained.

8.6.2.1 STRUCTURE OF CARBON NANOFIBER

The structural diversity of CNFs occurs by the anisotropic alignment of grapheme layers like the graphite, providing several particular structures such as platelet, herringbone and tubular CNFs according to the alignment direction to the fiber axis [87].

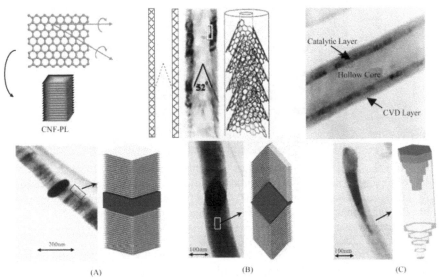

(A)   (B)   (C)

FIGURE 8.2   Different structures of carbon nanofibers (CNFs): (a) platelet structure, (b) herringbone structure, and (c) tubular structure.

There are mainly three types of carbon nanofibers: the herringbone in which the graphene layers are stacked obliquely with respect to the fiber axis; the platelet in which the graphene layers are perpendicular to the fiber axis; and the tubular in which the graphene layers are parallel to the growth axis [55].

Figure 8.3 shows the SEM image of carbon nanofiber film. It can be seen from Fig. 8.2a that the film is composed of aggregated nanofibers. They have been called *platelet carbon nanofibers*. The fibers are of several tens μm long, zigzag, and most of them have a bright ellipsoidal particle on their tip. The particle has the nearly same width as that of the PCNFs [88].

FIGURE 8.3   SEM images of platelet carbon nanofibers film. (a) SEM image of PCNFs film (b) Energy dispersive X-ray spectrometry (EDXS) of a light particles (indicated by _A_ in (a)) on the top of a fiber. (c) The size of a typical PCNFs, which shows the width (100–300 nm) and thick (30 nm) of the fiber (d) surface morphology of a PCNF.

Herringbone-type carbon nanofibers (CNFs) are a special kind of CNF with angles between the graphene plane direction and the axis of CNFs in the range of 0–90°. Theoretical calculations indicate that this type of nanofiber may exhibit some excellent properties, such as enhanced field emission, caused by the open edge sites, Aharonov–Bohm magnetic effects and magneto conductance, and localized states at the Fermi level may give rise to materials with novel electronic and magnetic properties Moreover, they can be expected to be used as absorbent materials, catalyst supports, gas storage materials and composite fillers, due to their special structural characteristics. Herringbone-type CNFs with large diameter and a very small or completely hollow core have been synthesized through a catalytic chemical vapor deposition (CVD) method [38].

8.7   POLYACRYLONITRILE BASED CARBON NANOFIBERS

As we discuss there are various methods to produce carbon nanofibers or carbon nanotubes, for example, vapor growth [89], arc discharge, laser ablation and chemical vapor deposition [28, 89]. However, these are very expensive processes owing to the low product yield and expensive equipment. Preparation carbon nanofiber by electrospinning of proper precursors is preferred because of its lower cost and more output [90].

Polyacrylonitrile is a common precursor of general carbon nanofibers. As described in the previous section, PAN is the most common Polymer for the preparation of CNFs due mainly to its relatively high melting point and carbon yield and the ease of obtaining stabilized products by forming a thermally stable structure. Additionally, the surface of PAN-based CNFs can be modified and functionalized using a coating or activation process [18]. Moreover, PAN can be blended with other polymers (miscible or immiscible) to carry out coelectrospinning or be embedded with nanoscale components (e.g., nanoparticles, nanowires, nanotubes or catalysts) to obtain multiphase precursors and subsequently to make composite CNFs through high-temperature treatment.

PAN nanofibers were produced using electrospinning by dissolving polyacrylonitrile in N, N-dimethylformamide (DMF) solution. The spinning can be carried out for 8, 10, 13, 15 & 20% (by weight) concentrations at different voltages, flow rates, varying distance between needle and collector and at different needle diameters [91]. Z. Kurban et al. have synthesized CNF by heat treatment of electrospun Polyacrylonitrile in dimethylsulphoxide, offering a new solution route of low toxicity to manufacture sub-60 nm diameter CNFs [28, 92].

After that PAN nanofiber would be stabilized and carbonized which will be discussed later. During the stabilization, PAN was stretched to improve the molecular orientation and the degree of crystallinity to enhance the mechanical properties of the fibers [85].

Chun et al. [93] produced carbon nano fibers with diameter in the range from 100 nm to a few microns from electrospun polyacrylonitrile and mesophase pitch precursor fibers. Wang et al. [94, 95] produced carbon nanofibers from carbonizing of electrospun PAN nanofibers and studied their structure and conductivity. Hou et al. [96] reported a method to use the carbonized electrospun PAN nanofibers as substrates for the formation of multiwall carbon Nanotubes. Kim et al. [14, 97] produced carbon nanofibers from PAN-based or pitch-based electrospun fibers and studied the electrochemical properties of carbon nanofibers web as an electrode for supercapacitor.

Sizhu wu et al. [85] electrospun PAN nanofibers from PAN solution in dimethylformamide (DMF) and hot-stretched them by weighing metal in a temperature controlled oven to improve its crystallinity and molecular orientation, and then were converted to carbon nanofibers by stabilization and carbonization. PAN was stretched to improve the molecular orientation and the degree of crystallinity to enhance the mechanical properties of the fibers. The hot-stretched nanofiber sheet can be used as a promising precursor to produce high-performance carbon nanofiber composites.

Several research efforts have been attempted to prepare the electrospun PAN precursor nanofibers in the form of an aligned nanofiber bundle [50, 98]. The bundle was then tightly wrapped onto a glass rod, so that tension existed in a certain degree during the oxidative stabilization in air.

TABLE 8.3   Polyacrylonitrile /DMF Solution Properties

Viscosity (Cp)	Temperature (°C)	Conductivity (µs)	Surface tension (m N/m)	Solution composition		Reference
				Polymer	DMF	
8%	333.3	20.8	39.0	72.9	8.0	[99]
10%	1723.0	22.0	43.2	77.0	10.0	[100]
13%	2800.0	21.4	50.0	97.6	13.0	[101]
15%	3240.4	23.3	56.7	105.0	15.0	[102]
20%	3500	21.1	58.0	122.5	20.0	[102]

The mechanical resonance method and Weibull statistical distribution can be used to analyze the mechanical properties (i.e., bending modulus and fracture strength) of CNFs. Using these methods, Zussman et al. [103] found that the stiffness and strength of CNFs were inferior to those of commercial PAN-based carbon fibers. The average bending modulus of individual CNFs was 63 GPa, lower than the lowest tensile modulus (230 GPa) for commercial High-strength PAN-based carbon fibers. Nevertheless, Zussman et al. [103] pointed out that the mechanical properties had the potential to be significantly improved if the microstructure of precursor nanofibers and thermal treatment process were optimized.

8.8   ELECTROSPINNING METHODOLOGY

Electrospinning is a process involving polymer science, applied physics, fluid mechanics, electrical, mechanical, chemical, material engineering and rheology [104]. It has been recognized as an efficient technique for the fabrication of fibers in nanometer to micron diameter range from polymer solutions or melts [105]. In a typical process, an electrical potential is applied between a droplet of polymer solution or melt held at the end of a capillary and a grounded collector. When the applied electric field overcomes the surface tension of the droplet, a charged jet of polymer solution or melt is ejected. The jet grows longer and thinner due to bending instability or splitting [83] until it solidifies or collects on the collector.

The fiber morphology is controlled by the experimental parameters and is dependent upon solution conductivity, concentration, viscosity, polymer molecular weight, applied voltage, etc. [34, 106] much work has been done on the effect of parameters on the electrospinning process and morphology of fibers.

Sinan Yördem et al. [107] reported that the fiber diameter increased with increasing polymer concentration according to a power law relationship. Filtering application [108] is also affected by the fiber size. Therefore, it is important to have control over the fiber diameter, which is a function of material and process parameters.

Deitzel et al. [109] reported a bimodal distribution of fiber diameter for fibers spun from higher concentration solution. Boland et al. [110] obtained a strong linear relationship between fiber diameter and concentration in electrospun poly(glycolic acid) (PGA). Ryu et al. [111] and Katti et al. [112] also reported a significant relationship between fiber diameter and concentration in electrospinning process. For the effect of applied voltage, Reneker et al. [113] obtained a result that fiber diameter did not change much with electric field when they studied the electrospinning behavior of polyethylene oxide. Mo et al. and Katti et al. [112] reported that fiber diameter tended to decrease with increasing electrospinning voltage, although the influence was not as great as that of polymer concentration. But Demir et al. [114] reported that fiber diameter increased with increasing electrospinning voltage when they electrospun polyurethane fibers.

Sukigara [115] studied the effect of electrospinning parameters (electric field, tip-to-collector distance and concentration) on the morphology and fiber diameter of regenerated silk from Bombyx mori using response surface methodology and concluded that the silk concentration was the most important parameter in producing uniform cylindrical fibers less than 100 nm in diameter.

In order to optimize and predict the morphology and average fiber diameter of electrospun PAN precursor, design of experiment was employed. Morphology of fibers and distribution of fiber diameter of PAN precursor were investigated varying concentration and applied voltage [116].

Another advantage of the electrospinning is that it can be used to produce a web structure [117]. When used as an electrode, it does not need a second processing step adding a binder and an electric conductor such as carbon black. Therefore, the webs from electrospinning have important advantages such as an ease of handling, an increase in the energy density due to large specific surface area, an improvement of the conductivity due to increased density of the contact points, and low cost of preparations of the electrodes [14].

During the electrospinning process, the droplet of solution at the capillary tip gradually elongates from a hemispherical shape to a conical shape or Taylor cone as the electric field is increased. A further increase in the electric field results in the ejection of a jet from the apex of the cone. The jet grows longer and thinner until it solidifies and collects on the collector. The fiber morphology is controlled by process parameters. In this study the effect of solution concentration and applied voltage on fiber morphology and average fiber diameter were investigated. Polymer concentration was found to be the most significant factor controlling the fiber diameter in the electrospinning process [78] (Table 8.4).

TABLE 8.4 The Electrospinning Conditions that Gave the Best Alignment of the Nanofibers [116]

Concentration (Wt%)	Tip to target distance (cm)	Voltage (kv)	Width of gap (cm)	Fiber diameter (nm)
10	15	9	1	165±25
11	15	9	1.2	205±25
12	20	9.5	1.5	245±20
13	20	10	2.2	255±30
14	20	11	2.5	300±30
15	20	11	3	400±40

8.9   ELECTROSPINNING OF PAN SOLUTION

The proper solvent has to be selected for dissolving a polymer source completely before carrying out the electrospinning. DMF is considered to be the best solvent among various organic solvents due to its proper boiling point (426K) and enough electrical conductivity (electrical conductivity = 10.90 mS/cm, dipole moment = 3.82 Debye) for electrospinning [1, 41, 95, 98]. This mixture should be vigorously stirred by an electromagnetically driven magnet at around 60 °C until it becomes a homogeneous polymer solution. Different concentrations of PAN solution (8–20% (wt)) can be used [91] (Fig. 8.4).

The following parameters affect the PAN solution electrospinning:

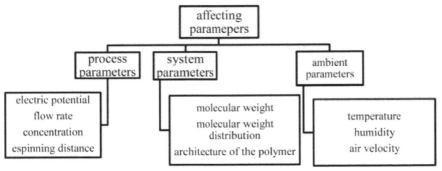

FIGURE 8.4   Classifications of parameters affect PAN solution electrospinning.

The electrospinning equipment for PAN solution can act with two different collectors, stationary and rotating drum. One important physical aspect of the electrospun PAN nanofibers collected on the grounded collectors is their dryness from the solvent used to dissolve the polymer, i.e., DMF. At the distances of tip to target of 5 and 7.5 cm, the structures of nanofibers were not completely stabilized and consequently the cross-sections of spun nanofibers became more flat and some nanofibers shuck together and bundles of nanofibers were collected. At the distances of tip to target of 15 cm and longer, the nanofibers exhibited a straight, cylindrical morphology indicating that the nanofibers are mostly dry when they have reached the target [118]. Another important factor is electrospinning distance, as we know a certain minimum value of the solution volume suspended at the end of the needle should be maintained in order to form an equilibrium Taylor cone. Therefore, different morphologies of electrospun nanofibers can be obtained with the change in feeding rates at a given voltage. Jalili et al. [118] proved that at lower feeding rate of 2 mL/h, a droplet of solution remains suspended at the end of the syringe needle and the electrospinning jet originates from a cone at the bottom of the droplet. The nanofibers produced under this condition have a uniform morphology and no bead defect was present. At lower feeding rate of 1 mL/h, the solution was removed from the needle tip by the electric forces, faster than the feeding rate of the solution onto the needle tip. This shift in the mass balance resulted in sustained but unstable jet and nanofibers with beads were formed (Fig. 8.5).

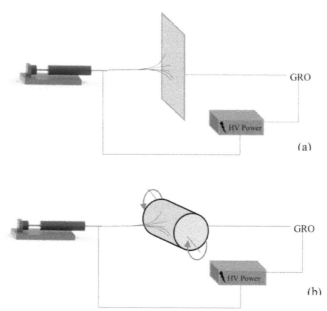

FIGURE 8.5 Electrospinning apparatus with (a) a stationary, grounded target; and (b) a rotating grounded target.

Studies confirmed that PAN nanofibers were formed by varying the solution concentration, voltage, solution flow rate and distance between needle and collector. The morphology of the PAN fibers for 10, 15, and 20% concentrations were studied using SEM and is shown in Fig. 8.6 (a–c), respectively [91]. As the concentration of the solution was increases, the diameter of the fiber was also increases.

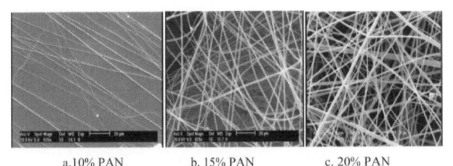

a.10% PAN   b. 15% PAN   c. 20% PAN
FIGURE 8.6   SEM analysis of PAN fibers for 10, 15 & 20% concentration.

Ma et al. [119] demonstrated a facile method for the preparation of porous ultrafine nanofibers of PAN as a precursor for carbon nanofiber. They prepared the PAN/NaHCO$_3$ composite nanofibers by electrospinning, and then NaHCO$_3$ was removed by a selective dissolution and reaction with the solution of hydrochloric acid (10 wt %). The obtained PAN fibers showed highly porous surfaces after the extraction of NaHCO$_3$.

For many applications, it is necessary to control the spatial orientation of 1D nanostructure. In the fabrication of electronic and photonic devices, for example, well aligned and highly ordered architectures are often required [120, 121]. Even for application as fiber-based reinforcement, it is also critical to control the alignment of fibers [50]. Because of the bending instability associated with a spinning jet, electrospun fibers are often deposited on the surface of collector as randomly oriented, nonwoven mats. The whipping instability is mainly caused by the electrostatic interactions between the external electric field and the surface charges on the jet. The formation of fibers with fine diameters is mainly achieved by the stretching and acceleration of the fluid filament in the instability region [98]. In the past several years, a number of approaches have been demonstrated to directly collect electrospun nanofibers as uniaxially aligned arrays [50, 98, 120–122]. The most popular method of obtaining aligned fibers is by using rotating drum target.

The aligned fiber mechanism behind the rotating drum technique is as follows:

When a linear speed of the rotating drum surface, which serves as a fiber take-up device, matches that of evaporated jet depositions, the fibers are taken up on the surface of the drum tightly in a circumferential manner, resulting in a fair alignment. Such a speed can be called as an *alignment speed.* If the surface speed of the cylinder is slower than the alignment speed, randomly deposited fibers will be collected, as it is the fast chaos motions of jets determine the final deposition manner. On the other hand, there must be a limit rotating speed above which continuous fibers cannot be collected since the over fast take-up speed will break the fiber jet. The reason why a perfect alignment is difficult to achieve can be attributed to the fact that the chaos motions of polymer jets are not likely to be consistent and are less controllable [91].

8.10   STABILIZATION

The next step in preparing carbon nanofiber is stabilization. In stabilization, in order to prevent precursor fibers fusing together during carbonization, the thermoplastic precursor nanofibers are converted to highly condense thermosetting fibers by complex chemical and physical reactions, such as dehydrogenation, cyclization and polymerization [54]. If the temperature is raised too rapidly during polymerization, a very large amount of heat will be released leading to a loss of the orientation and melting of the polymer. Therefore, the heating rate during stabilization is usually controlled at a relatively low value (e.g., 1 °C min^{-1}). It is noted that external tension is necessary during stabilization to avoid shrinkage of the fibers and to maintain the preferential orientation of the molecules along the fiber axis [41].

Stabilization process which is carried out in air (oxidative stabilization) constitutes the first and very important operation of the conversion of the PAN fiber precursor to carbon as well as activated carbon fiber. During stabilization, the precursor fiber is heated to a temperature in the range of 180–300 °C for over an hour. Because of the chemical reactions involved, cyclization, dehydrogenation, aromatization, and oxidation and cross linking occur and as a result of the conversion of C≡N bonds to C=N bonds fully aromatic cyclized ladder type structure forms [36, 49].

This new structure is thermally stable (infusible). Also, it has been reported that during stabilization, CH2 and CN groups disappear while C=C, C=N and = C–H groups form. At the same time the color of precursor fiber changes gradually and finally turns black when carbonized [36]. Research shows that optimum stabilization conditions lead to high modulus carbon fibers. Too low temperatures lead to slow reactions and incomplete stabilization, whereas too high temperatures can fuse or even burn the fibers [49, 93, 117, 123]

8.11 CARBONIZATION

Carbonization is the last step for producing carbon nanofiber [49, 124]. It involves cross linking, reorganization and the coalescence of cyclized sections accompanying the structural transformation from a ladder structure to a graphite-like one and a morphological change from smooth to wrinkle. This process needs to be carried out in a dynamic inert gas atmosphere (e.g., nitrogen and argon), which can prevent oxidation, remove the pyrolysis products (i.e., volatile molecules such as, H_2O, H_2, HCN, N_2, NH_3 and CO_2) and transfer energy. Vacuum carbonization can also be employed, but the degree of carbonization is lower than that in nitrogen or argon. Furthermore, fiber shrinkage occurs during carbonization. Fiber shrinkage may further lead to the formation of a large number of pits on the surface of CNFs having a deleterious effect on their appearance [41].

During this process, the noncarbon elements remove in the form of different gasses and the fibers shrink in diameter and lose approximately 50% of its weight [54].

Wang et al. [125] proved that the conductivity of PAN-based carbon nanofibers produced by electrospinning increases sharply with the pyrolysis temperature, and also increases considerably with pyrolysis time at lower pyrolysis temperatures.

The stabilized nanofiber would be subsequently carbonized at a temperature around 1000 °C in an inert (high purity nitrogen gas) environment with the heating rate 1–2°C/min [34, 50, 116]. Zhou, et al. reported that the carbonized PAN nanofiber bundles can be further carbonized in vacuum at relatively high temperatures between 1400 °C to 2200 °C. A Lindberg high temperature reactor with inside diameter and depth of 12 cm and 25 cm, respectively, can be used for conducting the high-temperature carbonization; and the heating rate must be set at 5 °C/min. All of the carbonized PAN nanofiber bundles must be held at the respective final temperatures for 1 h to allow the carbonization to complete [50].

The optimum processing conditions (i.e., temperature and time) for stabilization and carbonization can be investigated using TGA. Micro structural parameters for CNFs and ACNFs, such as integrated intensity ratio of D and G peaks (RI), interlayer spacing (d002), crystallite size, molecular orientation and pore characteristics, which are of great significance for final applications, can be analyzed with Fourier transform infrared (FTIR) spectroscopy, Raman spectroscopy, XRD, electron energy loss spectroscopy (EELS), transmission electron microscopy (TEM) and isothermal nitrogen adsorption/desorption [35, 49, 123]. The disappearance of peaks of functional groups in FTIR spectra and a very broad diffraction peak at a 2θ value of about 24° in XRD can demonstrate the conversion from the as-spun nanofibers to CNFs. The disappearance of peaks of functional groups in FTIR spectra and a very broad diffraction peak at a 2θ value of about 24° in XRD can demonstrate the conversion from the as-spun nanofibers to CNFs. The degree of carbonization can be evaluated with Raman spectroscopy, EELS and XRD. In Raman spectra, the lower the RI value, the higher the degree of the transformation from disordered carbon to graphitic carbon and the fewer the number of defects in the CNFs. The in plane graphitic crystallite size (La) can be further calculated from RI according to the equation La = 4.4/RI (nm) [90, 126, 127]. In EELS, the degree of carbonization can be monitored by analyzing the contents of sp2 bonds (structural order) and sp3 bonds (structural disorder). In XRD spectra, the average crystallite dimensions (e.g., Lc (002) and La (110)) and d002 can be calculated using the Scherrer and Bragg equations. It is noticed that micro structural parameters measured from XRD and Raman spectra are usually inconsistent due to the different measurement mechanisms; that is, XRD provides average bulk structural information, whereas Raman spectroscopy gives structural information merely within the surface layer (ca 10 nm). The commonly observed core–shell structure of CNFs can be clearly inspected using high-resolution TEM. The specific surface area and average pore size of CNFs can be measured based on the Brunner–Emmett–Teller (BET) method. Mesopore and micropore size distribution can be obtained using the Barrett–Joiner–Halenda (BJH) and Horvath–Kawazoe (HK) methods, respectively [41] (Fig. 8.7).

The morphology of as spun, stabilized and carbonized electrospun 15% PAN fibers are studied using scanning electron microscopy (Fig. 8.8 (a–c)) [91]. All these figures show the formation of random fibers. The diameter of the fiber decreases with increase in stabilization and carbonization temperatures.

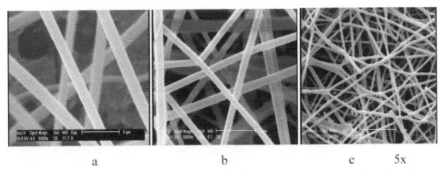

As spun (1-1.3µ)   Stabilized @ 280°C (0.9-1.1µ)   Carbonized @ 900°C

FIGURE 8.7 SEM analysis of as spun, stabilized and carbonized PAN fibers (magnification of 5000x).

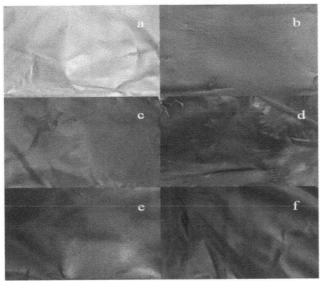

FIGURE 8.8 Stabilized samples in different temperatures a) 180, b) 200, c) 220, d) 250, e) 270 °C.

A problem during the heat treatment of the web of these nanofibers is that crack and wrinkle occur on a sheet only when holding on a metal plate because the nanofiber sheet cannot be shrunk. Using a fixing frame, which holds the sheet and prevents it from shrinkage can keep the structure safe [128] (Fig. 8.9).

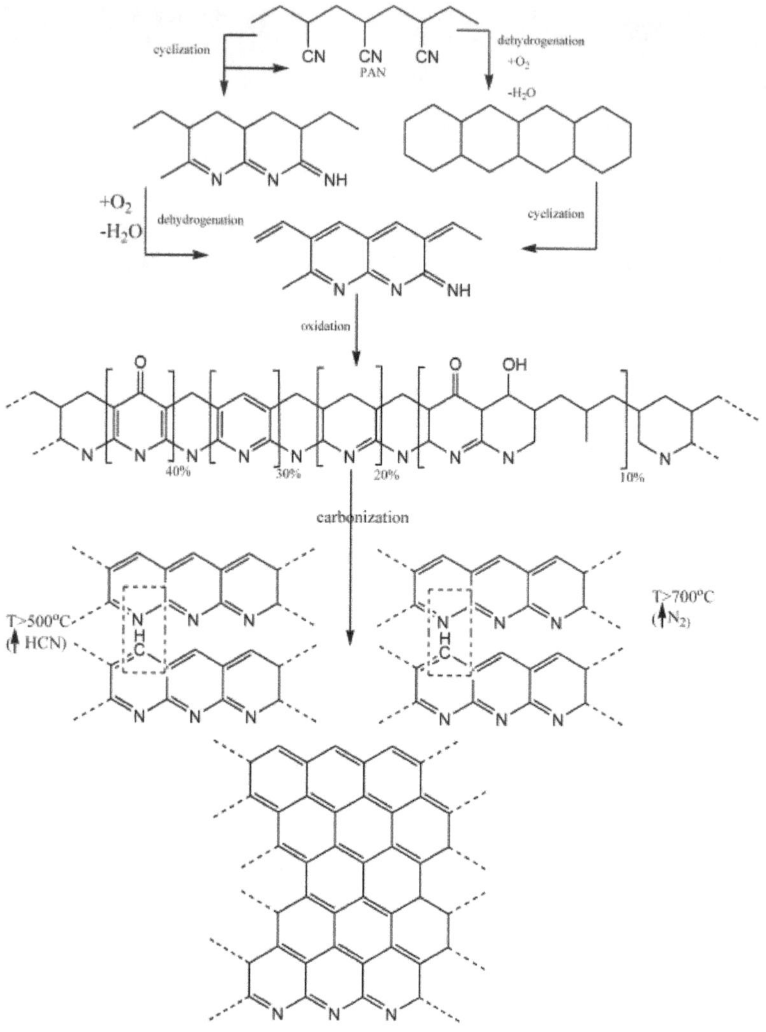

FIGURE 8.9   Process of PAN stabilization and subsequent carbonization.

8.12   COMPARISON OF STABILIZATION AND CARBONIZATION

The average diameter of the stabilized PAN nanofibers appeared to be almost the same as that of the as-electrospun nanofibers, while the average diameters of the carbonized PAN nanofibers were significantly reduced [14]. With increase of the final carbonization temperature, the carbon nanofibers became more graphitic and structurally ordered. The microstructure of the low-temperature (1000 °C) carbonized nanofibers was primarily turbostratic and the sheets of carbon atoms were folded and/or crumpled together, while the microstructure of the high-temperature (2200 °C) carbonized nanofibers was graphitic and the grapheme sheets stacked together to form ribbon-shaped structures [62].

Both electrical conductivities and mechanical properties of the carbon nanofiber increased with the increase of the final carbonization temperature. It is noteworthy that the electrical conductivities and mechanical properties of the carbon nanofiber cannot be directly interpreted as those of individual nanofibers in the bundles [35, 123].

In order to develop carbon nanofibers with superior mechanical properties particularly tensile strength, the electrospun PAN precursor nanofibers have to be extensively stretched; the stabilization (and probably carbonization as well) has to be conducted under optimal tension; and the PAN copolymer instead of homopolymer has to be used as the precursor because the electrospun carbon nanofibers with superior mechanical and electrical properties are expected to be an innovative type of nanomaterials with many potential applications [50]. Commercial PAN Precursor fibers are usually drawn prior to stabilization in order to reduce the probability of encountering a critical flaw during thermal treatment. However, few studies have applied stretching to electrospun precursor nanofibers before stabilization [129]. Additionally, although aligned nanofibers with various degrees of alignment have been obtained by using various modified collecting devices, tension is rarely applied to the nanofiber assembly during stabilization to prevent shrinkage of the fibers and to ensure molecular orientations along the fiber axis to a large extent. Therefore, there is still considerable room for further improvement of the micro structural, electrical and mechanical properties of the final CNFs. Considering the increased interest in electrospinning and the wide range of potential applications for electrospun CNFs, commercialized products can be expected in the future [41, 129].

8.13 FT-IR STUDY OF PAN AND STABILIZED PAN FIBER

IR spectra are considered as tool for determination the chemical interaction during heat treatment on PAN fibers. By using of these spectra, it is possible to study the relation between chemical changes and strength, aromatic index and fiber contraction during fabrication process. But analyzing these relations is so difficult because the intensity of bonds used for analysis depends on samples type, form and the way of preparation. The study of FT-IR spectra of PAN fibers sample with different co monomers shows that during stabilization of PAN fibers, the peaks related to C≡N bonds and CH_2 are reduced sharply. These reductions are related to cyclization of nitrile groups and stabilization procedure [54].

FT-IR spectra of PAN fibers have many peaks which related to existence of CH2, C≡N, C=O, C–O and C–H bonds. The absorption peaks are in range of 2926–2935 cm^{-1} are related to C–H bonds in CH, CH_2 and CH_3 but in this range the second weak peak is observed which is related to C-H bonds also. Another peak is observed in the range of 2243–2246 cm^{-1}, which is related to presence nitrile (C≡N) bonds and indicates the nitrile group exists in polyacrylonitrile chain. The absorption peaks in the ranges of 1730–1737 cm^{-1} and 1170 cm^{-1} are related to C=O or C–O bonds and are resulted from presence of comonomers like MA. Absorption in the range of 1593–1628 cm^{-1} is related to resonance C–O bonds. The peaks in the range of 1455–1460 cm^{-1} is related to tensile vibration and peaks in the range of 1362–1382 cm^{-1} and range of 1219 cm^{-1} are related to vibration in different situation [54, 130].

After stabilization, links in the range of 2926–2935 cm^{-1} are reduced which are related to CH_2 bond. These links are weakened and with some displacement are observed in the range of 2921–2923 cm^{-1}. Additionally these bonds in the range of 1455–1460 cm^{-1} are mainly omitted or would be reduced which is related to CH_2 bonds too. Also, the main reduction is observed in links in the range of 2243–2246 cm^{-1} which is resulted from the change C≡N bonds and their conversion to C=N. C≡N peak is weakened in stabilized samples or completely would be removed [54, 131]. New peaks in the range of 793–796 cm^{-1} are created as the result of =C–H bond formation. Increasing the intensity of =C–H groups and reduction of intensity of CH_2 groups shows that =C–H is created during aromatization of structure in the presence of oxygen.

During the stabilization step, the absorption peaks in the ranges of 1730–1737 cm^{-1} and 1170 cm^{-1} (related to C=O or C–O bonds) and two groups of peaks in the ranges of 1362–1382 cm^{-1} and 1219 cm^{-1} (related to C–H in different situation) mainly are removed in types of fibers. In the range of 670 cm^{-1}, some changes are observed which are not related to chemical changes of PAN fibers during oxidized stabilization. It is believe that this issue is may be related to different modes of CN and C–CN bonds appearance in FT-IR [130].

8.14   CNF FROM OTHER TYPE OF PRECURSORS

Many investigations shave dealt with PAN-based CNFs, as presented above. However, other precursors (e.g., pitch, PVA, PI and PBI) have also been used successfully to prepare electrospun CNFs. Pitch is generally obtained from petroleum asphalt, coal tar and poly(vinyl chloride) with a lower cost and a higher carbon yield compared with PAN [41]. Pitch is generally obtained from petroleum asphalt, coal tar and poly (vinyl chloride) with a lower cost and a higher carbon yield compared with PAN. However, impurities in the pitch are required to be fully removed to obtain high-performance carbon fibers leading to a great increase in the cost. Therefore, there have been limited studies on electrospun pitch-based carbon microfibers [93].

The diameter of the electrospun pitch fibers was in the micrometer range and difficult to become thinner due to the low boiling point (65–67 °C) of the solvent tetrahydrofuran (THF). Electrical conductivity was increased on increasing the carbonization temperature from 0 (700 °C) to 83 S cm^{-1} (1200 °C).

PVA, another thermoplastic precursor, allows the preparation of CNFs through thermal treatment processes [132]. Unlike thermoplastic precursors, electrospun thermosetting nanofibers can directly undergo carbonization for preparing CNFs without the need of the costly stabilization process. Therefore, they have gained increasing attention in recent years. For example, electrospun PI-based CNFs were prepared by Kim et al. The electrical conductivity of the CNFs carbonized at 1000 °C was measured to be 2.5 S cm^{-1}, higher than that (1.96 S cm^{-1}) of PAN-based CNFs treated at the same carbonization temperature [90]. PBI and PXTC are other precursors for CNF preparation [41].

Another thermosetting polymer precursor (PXTC) in the form of aligned nanofiber yarn (parallel with the electric field lines), which was several centimeters long, was obtained by electrospinning merely using a conventional flat aluminum plate as the collector. The formation of the yarn might result from the ionic conduction of PXTC. The presence of

D and G peaks in the Raman spectra demonstrated a successful conversion from electrospun PXTC yarn to CNFs in the temperature range from 600 to 1000 °C, while the yarn carbonized at 500 °C did not show these two peaks in the Raman spectra [41]. The mole fraction of graphite for the carbonized nanofibers was determined to be 0.21–0.24, less than that (0.25–0.37) obtained by Wang et al. [126] for PAN-based CNFs treated from 873 to 1473K, showing a lower degree of carbonization. Depending on the properties of the precursor nanofibers and subsequent high-temperature treatment processes, diverse microstructural and electrical properties have been achieved [41].

8.15   ACTIVATION

The most principal disadvantage of CNFs is its relatively low surface area and porosity (around 10–200 m²/g), which limit the applications of these materials like hydrogen storage or catalyst support; therefore is necessary increase the surface area to improve the yield in these materials. It is conceivable that CNF webs consisting of CNFs with porous surfaces, which are obtained through the activation of electrospun PAN-based CNFs, can lead to a significant expansion of applications of CNFs, such as in electrode materials, high-temperature filtration and removal of toxic gases.

The surface area can be modifying by means of activation process in which a part of structural carbon atoms are eliminated (mainly, the most reactive) by an activate agent. As consequence, the porosity and surface area increase and so, their applications as hydrogen storage or catalyst support improve [55].

Barranco et al. [86] expressed that Activation does not produce any important change in the shape, surface roughness, diameter, graphene sheet size, and electrical conductivity of starting nanofibers; it leads to new micropores and larger surface areas as well as a higher content of basic oxygen groups.

During the activation treatment, high porosity is formed within the material through the interaction of activating agents (usually oxidizing medium like steam, carbon dioxide, etc.) with carbon structures. By this interaction, the surface chemical properties are altered or transformed to some extent [133].

It is known there are still some amounts of nitrogen left in the structures of carbon fibers, which exist in various types of nitrogen functionalities after the carbonization treatment. These functionalities were bounded or attached to the edge parts of the carbon structures. During activation treatment, these edge parts would be preferentially attacked by the oxidizing agents, which probably led to the removal of these edge structures. Due to this effect, the surface nitrogen level decreased upon activation treatment. Wang et al. suggested that the activation treatment in steam helps the elimination of nitrogen from the carbon structure [133].

Comparatively to nanotubes, nanofibers present a nanostructure made of grapheme layer stacking which is favorable to activation. Two activation systems can be used for activated carbon nanofibers: physical activation by CO_2 or heat, and chemical activation by KOH or RbOH [134, 135]. A range of potential adsorbents was thus prepared by varying the temperature and time of activation. The structure of the CNF proved more suitable to activation by KOH than by CO_2, with the former yielding higher surface area carbons (up to 1000 m^2 g^{-1}).

Depending on the final application of the activated materials, it is possible to control their pore structure by choosing the suitable activation conditions. The increased surface area, however, did not correspond directly with a proportional increase in hydrogen adsorption capacity. Although high surface areas are important for hydrogen storage by adsorption on solids, it would appear that it is essential that not only the physical, but also the chemical, properties of the adsorbents have to be considered in the quest for carbon based materials, with high hydrogen storage capacities [106, 136].

8.15.1 PHYSICAL ACTIVATION

Physical activation involves carbonization of a carbonaceous precursor followed by gasification of the resulting char or direct activation of the starting material in the presence of an activating agent such as CO_2, steam or a combination of both, that among them steam activation is more common. This gasification or activation process eliminates selectively the most reactive carbon atoms of the structure generating the porosity [56, 129].

Steam is widely used as an activating agent for fabricating ACNFs because of its low cost and environmentally friendly character. Kim and

Yang [14] reported that the specific surface area of steam-activated carbon nanofibers (steam-ACNFs) decreased with increasing activation temperature (from 700 to 850 °C) due to the unification of micropores at elevated temperatures, whereas the electrical conductivity of ACNF electrodes and the accessibility of ions were increased according to cyclic voltammetry curves and impedance Nyquist plots.

For physical activation, about 2 g of CNF are placed in the center of a quartz tube in a tube furnace. Then the CNF are heated to the required reaction temperature (800–1000 °C) under neutral environment, for a reaction time of 15–45 min. The sample is then allowed to cool under argon [106]. The ACNFs activated at 800 °C afforded the highest specific surface area but low mesopore volume [137]. After activation, the average diameters of the fibers decreased about 100 nm compared with the CNFs without activation process. The reduction in diameter may be a consequence of the subsequent carbonization and burn-off through activation at elevated temperatures. However, no severe shrinkage is found in the ACNFs. Therefore, activation does not create defects on the surfaces; although the roughness of the ACNF surfaces is somewhat increased and nano-sized fibers are created [137] (Fig. 8.10).

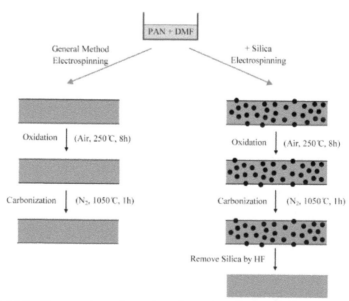

FIGURE 8.10 The procedure of manufacturing carbon fibers and silica-activated carbon fibers.

A representative case of physical activation is activation with silica, reported by Im et al. [138]. The activation agent was embedded into the fibers and then removed by physically removing the agents. The process of activation by silica is presented in Fig 8.11. The silica-activated carbons nanofibers are shown in Fig. 8.12. The pores generated by physical activation are clearly observed.

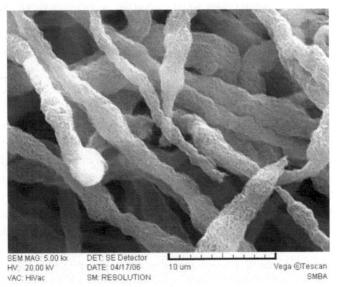

SEM MAG: 5.00 kx DET: SE Detector
HV: 20.00 kV DATE: 04/17/06 10 um Vega ©Tescan
VAC: HiVac SM: RESOLUTION SMBA

FIGURE 8.11 Silica-activated carbon nanofibers [138].

8.15.2 CHEMICAL ACTIVATION

In addition to the steam activation, activating agents (e.g., KOH, $ZnCl_2$, NaOH, Na_2CO_3, K_2CO_3, SiO_2) can also be used for activation of CNFs [41]. In chemical activation the precursor is impregnated with a given chemical agent and, after that, is pyrolyzed. As a result of the pyrolysis process, a much richer carbon content material with a much more ordered structure is produced, and once the chemical agent is eliminated after the heat treatment, the porosity is so much developed. Several activating agents have been reported for the chemical activation process: phosphoric acid, zinc chloride and alkaline metal compounds [42, 55, 56, 60, 106, 134–136, 139].

Phosphoric acid and zinc chloride are activating agents usually used for the activation of lignocellulosic materials which of coal precursors or chars have not been previously carbonized. Contrarily, alkaline metal compounds, usually KOH, are used for the activation (Table 8.5).

TABLE 8.5   Surface Area and Yield to Different Activating Agents [55]

Activating agents	Initial CNFS	KOH	NaOH	K_2CO_3	$KHCO_3$	$Mg(OH)_2$
Surface area (m²/g)	127.0	407.5	177.0	261.0	233.0	173.8
Yield (%)	-	55.1	76.2	62.7	67.2	77.5

An important advantage of chemical activation is that the process normally takes place at a lower temperature and shorter time than those used in physical activation. In addition, it allows us to obtain very high surface area activated carbons. Moreover, the yields of carbon in chemical activation are usually higher than in physical activation because the chemical agents used are substances with dehydrogenation properties that inhibit formation of tar and reduce the production of other volatile products [42, 134].

However, the general mechanism for the chemical activation is not so well understood as for the physical activation. Other disadvantages of chemical activation process are the need of an important washing step because of the incorporation of impurities coming from the activating agent, which may affect the chemical properties of the activated carbon and the corrosiveness of the chemical activation process [56].

Im et al. [36] studied the hydrogen adsorption capacities of ACNFs activated using KOH (chemical activation; applied after carbonization) and $ZnCl_2$ (physical activation; applied along with carbonization) with the volumetric method. Although KOH-activated CNFs had higher specific surface area and total volume than ZnCl2-activated CNFs, the ultra micropore (0.6–0.7 nm) volume of ZC-W4 (PAN:DMF:ZnCl2 = 3:50:4 weight ratio) or ZC-W6 (PAN:DMF:$ZnCl_2$ = 3:60:6 weight ratio) was much larger than that of KOH-activated CNFs. ZC-W6 with the largest ultra microspore volume (0.084 cm³ g⁻¹) exhibited the highest hydrogen adsorption capacity. Therefore, it was concluded that hydrogen adsorption capacity is mainly determined by ultra micropore (0.6–0.7 nm) volume. Beaded nanofibers are generally unfavorable during the electrospinning process.

It is known that the traditional activation processes are relatively complex and costly. To resolve this issue, it is possible to fabricate porous CNFs by precreating pores in the as-spun precursor nanofibers, such as by choosing a particular solvent system, by changing the environmental humidity or by using polymer mixtures, and then using a thermal treatment.

Cheng et al. produced chemically activated carbon nanofibers based on a novel solvent-free co extrusion and melt-spinning of polypropylene-based core/sheath polymer blends and their morphological and microstructure characteristics analyzed by scanning electron microscopy, atomic force microscopy (AFM), Raman spectroscopy, and X-ray diffractometry [139].

Kim et al. used the third strategy via the removal of a poly (methyl methacrylate) (PMMA) component during the carbonization of electrospun immiscible polymers (PAN and PMMA). The higher the PAN content, the finer the electrospun composite nanofibers. The carbon surface was burned off by generating carbon monoxide and carbon dioxide from outside to inside by the following reactions (1–5) (here, M = Na or K) [140]:

$$6 \text{ MOH} + C \leftrightarrow 2 \text{ M} + 3H2 + 2 \text{ M2CO3} - \tag{1}$$

$$\text{M2CO3} + C \leftrightarrow M2O + 2CO \tag{2}$$

$$\text{M2CO3} \leftrightarrow M2O + CO_2 - \tag{3}$$

$$2 \text{ M} + CO_2 \leftrightarrow M2O + CO \tag{4}$$

$$M2O + C \leftrightarrow 2 \text{ M} + CO \tag{5}$$

TABLE 8.6 Comparisons of Preparation Processes for Electrospun CNFs and ACNFs Discussed in the Open Literature

Precursor	Stabilization	Carbonization	Activation	Ref.
PAN/DMF-ZnCl2	250 °C (1h)	1050 °C(2h)	–	[36]
PAN/DMF-KOH	250 °C (1h)	–	750 °C(3h)	[36]
PAN/DMF- si	280 °C (5 h)	700 °C (1h)	–	[141]

TABLE 8.6   *(Continued)*

Precursor	Stabilization	Carbonization	Activation	Ref.
PAN/DMF	350 °C (0.5h)	750 °C(1h), 1100 °C(1h)	–	[117]
PAN/DMF	280 °C (1h)	1000 °C(1h)	–	[142]
PAN/DMF	280 °C (1h)	700, 800, 900 and 1000 °C (1h)	–	[90]
PAN/DMF-CNT	285 °C (2, 4, 8, 16 h)	700 °C (5 °C min^{-1}), 1 h))	900 °C (in CO_2 for 1h)	[51]
PAN/DMF-Mg	280 °C (1h)	1000 °C (at a rate of 5 °C/min^{-1}, 1h)	800 °C (0.5h)	[143]
PAN/DMF	280 °C (1h)	900 °C(1h)	–	[91]
PAN/DMF	240 °C (2h)	1200 °C (10 min)	–	[54]
PAN-PVP/DMF	270 °C (1h)	1000 °C 5 °C/min^{-1})	–	[1]
PAN/DMF	Not given	1200 °C (0.5 h)	–	[126]
PAN/DMF	280 °C (3h)	1000 °C (2 °C/min 1 h), 1400 °C 1800 °C, 2200 °C (1 h)	–	[50]
PAN/DMSO	125–283 °C (50 min)	1003–1350 °C	–	[49]
PAN/DMF	230 °C (2h)	600 °C(0.5h)	800, 850, and 900 °C (in CO_2 for 1 h)	[136]

8.16   APPLICATIONS

Many promising applications of electrospun activated carbon nanofibers can be expected if appropriate micro structural, mechanical and electrical properties become available.

8.16.1 HYDROGEN STORAGE

Due to the exhaustion of gasoline or diesel fuel, new energy sources are necessary to be developed as an assistant or alternative energy. Among them, hydrogen gas is an attractive possibility to provide new solutions for ecological and power problem [144]. Hydrogen is least polluting fuel. Since it is difficult to store hydrogen, its use as a fuel has been limited [145–147]. The possibility of developing hydrogen into an environmentally friendly, convenient fuel for transportation has lead to the search for suitable materials for its storage. The suitable media for hydrogen storage have to be light, industrial, and in compliance with national and international safety laws. In the last few years, researchers have paid much attention on hydrogen adsorption storage in nanostructured carbon materials [5], such as activated carbon [148], carbon nanotubes [146], and carbon nanofibers. Carbon nanomaterials due to their high porosity and large surface area have been suggested as a promising material for hydrogen storage [6] (Table 8.7).

TABLE 8.7 Summary of Reported Hydrogen Storage Capacity of Carbon Nanofiber

Sample	Purity	T(K)	P(Mpa)	H_2 (Wt %)	Reference
GNF	–	77–300	O.8–1.8	0.08	Ahn et al. [145]
GNF herringbone		298	11.35	67.58	Browning et al. [149]
GNF platelet		298	11.35	53.68	Browning et al. [149]
Vapor grown carbon fiber		298	3.6	<0.1	Tibbetts et al. [146]
CNF		77	12	12.38	Rzepka et al. [150]
ACNF with CO_2		300	11	0.33	Blackman et al. [106]
ACNF with KOH		300	10	0.42	Blackman et al. [106]

Rodriguez and Baker investigated that the hydrogen storage capacities of CNF at room temperature and pressures up to 140 bars were quantified independently by gravimetric and volumetric methods, respectively [145, 149, 150]. Ji Sun Im and Soo-Jin Park [36] studied the relation between pore structure and the capacity of hydrogen adsorption, textural properties of activated CNFs with micropore size distribution, specific surface area, and total pore volume by using BET (Brunauer–Emmett–Teller) surface analyzer apparatus and the capacity of hydrogen adsorption was evaluated by PCT (pressure–composition–temperature) hydrogen adsorption analyzer apparatus with volumetric method.

They indicated that Even though specific surface area and total pore volume were important factors for increasing the capacity of hydrogen adsorption, the pore volume which has pore width (0.6–0.7 nm) was a much more effective factor than specific surface area and pore volume in PAN-based electrospun activated CNFs.

Chemically activated carbon nanofiber with NaOH and KOH, were evaluated by Figueroa-Torres [5] for hydrogen adsorption at 77K and atmospheric pressure. Hydrogen adsorption reached values in the order of 2.7 wt% for KOH activated carbon. The mechanism of formation of the porous nanostructures was found to be the key factor in controlling the hydrogen adsorption capacity of chemically activated carbon [134].

Loading of CNF with metallic particles can enhance the hydrogen storage capacity of it. The hydrogen storage behaviors of porous carbon nanofibers decorated by Pt nanoparticles were investigated, It was found that amount of hydrogen stored increased with increasing Pt content to 3.4 mass%, and then decreased [151].

Fuel cells are used to convert hydrogen, or hydrogen-containing fuels, directly into electrical energy plus heat through the electrochemical reaction of hydrogen and oxygen into water. The process is that of electrolysis in reverse. Overall reaction [152]:

$$2 \, H_2 \, (gas) + O_2 \, (gas) \rightarrow 2 \, H_2O + energy \qquad (6)$$

Because hydrogen and oxygen gases are electrochemically converted into water, fuel cells have many advantages over heat engines. A simple example of a hydrogen fuel cell consists of an anode and a cathode with an electrolyte in-between that allows positive ions to pass through. Hydrogen fuel is fed to the anode and atmospheric oxygen is fed to the cathode.

When activated by a catalyst, usually platinum on the cathode itself, the hydrogen atoms separate into electrons and protons, which take different paths to the cathode. The electrons take a path through an electrical circuit and load, while the protons take a path through the electrolyte. When the electrons and protons meet again at the cathode, they recombine along with the oxygen atoms to produce water and heat. This process is illustrated in Figure below [153].

In order for a fuel cell to operate, it requires a constant supply of hydrogen. In transportation sector applications, this hydrogen must be stored locally in a safe and efficient manner. One method to do this that is proving to be viable and able to meet the D.O.E. benchmark is that of a pressurized tank containing hydrogen physically adsorbed on carbon [153]. Hydrogen has a kinetic diameter of 0.289 nm, which is slightly smaller than the ~0.335–0.342 nm interlayer spacing in carbon nanofibers (see Fig. 8.12.) [149].

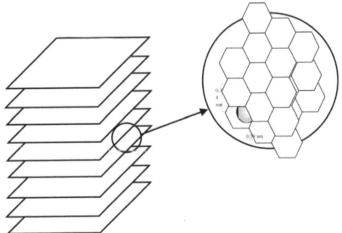

FIGURE 8.12   (a) Schematic representation of the structure of a carbon nanofiber; (b) enlarged section.

The principle application of a carbon nanofiber hydrogen storage medium is in a fuel tank for an integrated on-board fuel cell system with a polymer electrolyte membrane (PEM) fuel cell stack at its core and a hydrogen supply stored as adsorbed hydrogen in a pressurized tank containing carbon nanofibers.

In a PEM fuel cell, two half-cell reactions take place simultaneously, an oxidation reaction (loss of electrons) at the anode and a reduction reaction (gain of electrons) at the cathode. These two reactions make up the total oxidation-reduction (redox) reaction of the fuel cell, the formation of water from hydrogen and oxygen gases [152]. The D.O.E.'s target benchmarks for on-board hydrogen storage are based on a model hydrogen fuel cell powered vehicle to be able to travel 500 km without refueling and the metric that 3.1 kg of hydrogen would be required for a fuel cell powered car to travel those 500 km. Based on the 10–15wt% storage capability that has been demonstrated with properly prepared carbon nanofibers, as illustrated earlier in this paper, this would result in a 10 full tank of hydrogen adsorbed carbon nanofibers weighing between 21 and 31 kg with perhaps some additional weight required for the pressure valve and protective covering [154].

A new group of fuel cell is microbial fuel cells (MFCs), which is a novel technology that produces electricity using bacteria as electrocatalysts. The performance of MFCs is influenced by the type of electrode, the electrode distance, the type and surface area of their membrane, their substrate and their microorganisms. The most common catalyst used in cathodes is platinum (Pt). Ghasemi et al. applied chemically and physically activated carbon nanofibers as an alternative cathode catalyst to platinum in a two-chamber microbial fuel cell for the first time [155].

The main reason suggested for improved hydrogen adsorption was that electrospun activated carbon nanofibers might be expected to have an optimized pore structure with controlled pore size. This result may come from the fact that the diameters of electrospun fibers can be controlled easily, and optimized pore sizes can be obtained with a highly developed pore structure. To find the optimized activation conditions, carbon nanofibers were activated based on varying the chemical activation agents, reaction time, reaction temperature, and the rate of inert gas flow [156].

Porous carbon-nanofiber-supported nickel nanoparticles also can be used as a promising material for hydrogen storage. It was found that the amount of hydrogen stored was enhanced by increasing nickel content [7] (Fig. 8.13).

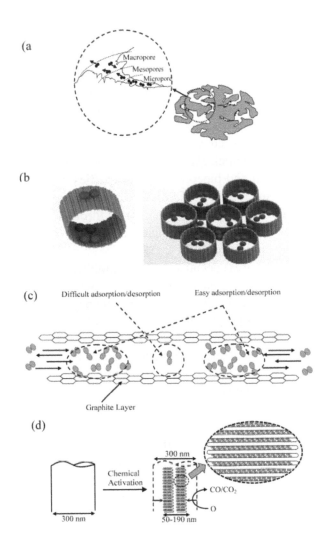

FIGURE 8.13 The mechanism of hydrogen adsorption using various carbon materials; (a) activated carbon, (b) single walled carbon nanotube, (c) graphite, (d) electrospun activated carbon nanofibers.

8.16.2 MEDICAL APPLICATIONS

CNTs and CNFs play an important role in nanomaterial research due to their mechanical, optical, electrical and structural properties. In the field of regenerative medicine, these nanofibers are becoming increasingly attractive as they can be modified to be integrated into human bodies for promoting tissue regeneration and treatment of various diseases [157]. Considering the excellent mechanical strength of CNTs and CNFs, it is natural that there are many studies focusing on using these carbon nanostructures as reinforcing agents in composite materials, especially in bone scaffolds. Since natural neural tissues have numerous nanostructured features (such as nanostructured extracellular matrices that neural cells interact with), CNFs/CNTs which also have such nanofeatures and exceptional electrical, mechanical and biocompatible properties, are excellent candidates for neural tissue repair. CNFs have excellent properties comparable to CNTs but at a lower cost and are fabricated through an easier scale-up process thus, CNFs have generated much interest in regenerative neural tissue engineering applications [157, 158].

Carbon nanostructures specially carbon nanofibers provide a large surface area with high surface energy which can easily increase the interaction of nano scaffolds and cells and improve the performance of implant, this fact make them appropriate for nerve regeneration [159]. CNF and ACNF can be applied in drug delivery system too. Drug delivery is an emerging field focused on targeting drugs or genes to a desirable group of cells. The goal of this targeted delivery is to transport a proper amount of drugs to the desirable sites (such as tumors, diseased tissues, etc.) while minimizing unwanted side effects of the drugs on other tissues. CNFs with the ability to cross cell membranes are good candidates to serve as drug delivery carriers to cells with high efficiency [157].

Some methods were used to modify CNFs to improve their biocompatibility properties and highlight some applications of these fibrous materials in creating regenerative scaffolds and drug and gene delivery vehicles [105, 110, 117]. Despite the tremendous potential CNFs can bring, toxicity of these materials is one of the issues that remain to be fully studied. The human health hazards associated with exposure to carbon nanoparticles have not been fully investigated, especially their potential for genotoxicity and carcinogenicity. Surprisingly, despite the current widespread use of carbon nanofibers, toxicological studies have mainly focused on carbon nanotubes, and only a few studies have evaluated different carbon nanofibers and their toxicity [79]. Importantly, the presence of unreacted catalysts in CNFs is a key factor promoting their toxicity so care should be taken when synthesizing CNFs. Clearly, toxicity effects of these fibers when implanted or injected need further investigation. With continued work from researchers, there is no doubt that these materials will become useful and safe to use for enhancing human health [108].

8.16.3 ENERGY STORAGE

Electrospun porous carbon nanofiber webs have attracted considerable attention as a promising electrode material in energy storage devices due to high electrical conductivity, high specific surface area and freestanding nature [160]. These materials possess some rather unique properties that may find use in a number of electrochemical energy storage systems including primary and secondary batteries, fuel cells and electrochemical capacitors. The structure and properties of the nanofibers can be controlled at the nanometer level by manipulating the process variables [2, 161, 162]. The application of CNF and activated form of it in energy storage, divides into different parts such as lithium-ion batteries and electrochemical double-layer capacitors.

8.16.3.1 CNF AND ACNF APPLICATION IN THIN AND FLEXIBLE LITHIUM-ION BATTERIES (LIBS)

In most batteries, porous structure is an essential requirement. A sponge-like electrode will have high discharge current and capacity, and a porous separator between the electrodes can effectively stop the short circuit, but allow the exchange of ions freely. Solid electrolytes used in portable batteries, such as lithium ion battery (LIB), are typically composed of a gel or porous host to retain the liquid electrolyte inside. A porous membrane with well-interconnected pores, suitable mechanical strength and high electrochemical stability could be a potential candidate [163]. Lithium ion batteries offer very high energy densities and design flexibilities, thereby making them integral in modern day consumer devices such as cellular phones, camcorders and laptop computers. However, unlike electrochemical capacitors, lithium ion batteries are restricted to low achievable power densities. With the advent of electric vehicles (EV) and plug-in hybrid electric vehicles (PHEV) there has therefore been a growing need to build lithium-ion batteries that can not only provide high energy densities but also deliver high power densities in order to be considered as a potential replacement for conventional gasoline engines [164]. A lithium ion battery essentially comprises of three components cathode, anode and electrolyte. Cathodes are generally categorized into three types, namely (1) lithium based metal oxides such as $LiCoO_2$, (2) transition metal phosphates such as Li3 V2 $(PO_4)_3$, and (3) spinels such as $LiMn_2O_4$. Among anodes, carbon is the typical material used in lithium ion batteries [164]. A lithium ion battery and its charging and discharging processes are depicted in Scheme 5 (Fig. 8.14).

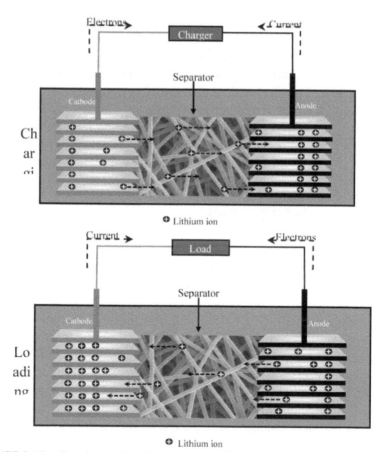

FIGURE 8.14   Charging and loading processes of lithium ion battery.

Nanostructured materials are attractive for lithium-ion batteries with their unique features that arise from their nanoscale structures. Due to the small size, the optimum transport of both electrons from the back contact to the front of the electrode and ions from the electrolyte to the electrode particles can lead to a rapid discharging and charging rate. Up to now, various nanomaterials such as metal oxides, carbonaceous materials, phosphate, and sulfides, etc., have been widely used as anode material for lithium-ion battery. Among the families of materials studied, nanostructured carbon materials such as carbon nanofibers have a lot of advantages such as availability, chemical stability, good cyclability, and low cost as anode materials for lithium-ion batteries [20].

These increases are attributed to the uniform distribution of network-like CNF and ACNF of high conductivity; CNF not only connects the surface of the active materials, its network penetrates into and connects each active material particle. CNF composite electrode also improves the electrochemical performance of thin and flexible lithium-ion batteries such as discharge capacity at high current densities, cycle-life stability, and low-temperature (at 20 °C) discharge capacity [20]. These improved electrochemical properties are attributed to the well-distributed network-like carbon nanofibers, within the cathode. The addition of CNF reduces the electron conducting resistance and decreases the diffusion path for lithium ions, hence increases the utilization of active materials during high-current discharge and low-temperature discharge. In addition, network-like CNF forms a more uniform cathode structure so as to have a lower deterioration rate and correspondingly better life cycle stability.

In order to further improve the performance of carbon materials as anodes for LIBs, an effective porous structure in a controllable fashion is needed to provide desirable surface area and open pore structure, which can achieve larger energy conversion density, higher rate capability, and better cycle performance. Therefore, activated or porous CNFs with large specific surface area and controlled pore structure could be an ideal candidate to meet these requirements [97].

Mn-based oxide-loaded porous carbon nanofiber anodes, exhibiting large reversible capacity, excellent capacity retention, and good rate capability, are fabricated by carbonizing electrospun polymer/Mn $(CH_3COO)_2$ composite nanofibers without adding any polymer binder or electronic conductor. The excellent electrochemical performance of these organic/inorganic nanocomposites is a result of the unique combinative effects of nano-sized Mn-based oxides and carbon matrices as well as the highly developed porous composite nanofiber structure, which make them prom-

ising anode candidates for high-performance rechargeable lithium-ion batteries [93].

8.16.3.2   CNF AND ACNF APPLICATIONS IN ELECTROCHEMICAL DOUBLE-LAYER CAPACITORS (EDLCS)

In addition to lithium ion battery, electrospun carbon nanofibers could be
used in electrochemical double-layer capacitors (EDLC). EDLCs or super
capacitors are promising high-power energy sources for many different
applications where high-power density, high cycle efficiency and long
cycle life are needed. In 1879, Helmholtz suggested that EDLCs accu-
mulate electrical energy that is generated by the formation of an electrochemical double layer suggested, at the interface between electrode and
electrolyte (non-Faradaic process), unlike secondary batteries such as the
lithium ion battery or nickel metal hydride battery, which are based on a
redox reaction (Faradaic process). This energy storage system based on a
non-Faradaic process provides very fast charge and discharge making the
EDLCs with the best candidates to meet the demand for high power and
long durability [165].

However, because the energy density of EDLCs is small compared to
that of rechargeable batteries, it is necessary to increase the capacitance
of EDLCs. Recently, the relationships between the porous structures and
electrochemical behavior have become increasingly important. Although
the use of various materials as EDLCs has been investigated, the application is limited in terms of specific energy. To enhance the specific energy
and power of EDLCs, several researchers have put much effort into the
development and modification of carbonaceous materials, such as controlling the pore size distribution, introducing electro-active metallic particles
or electro-conducting polymers, and fabricating hybrid type cells [166].
The double layer capacitance is correlated with the morphological properties of the porous electrodes, particularly the surface and the pore size distribution of the carbon materials. Thus, tailoring the porous structures of
carbon materials is a major goal of EDLC optimization [17, 44]. Various
forms of carbonaceous materials, i.e., powder, fiber, paper or cloth (fabric
or web), carbon nanotubes, carbon nanofibers, and related nanocomposites
are candidates for electrodes of EDLCs. A paper type material particularly

useful for application as electrodes as the addition of binder, which normally degrades the performance of capacitors, is not needed.

Electronic properties make the carbon nanostructures such as carbon nanofiber applicable inter alia in EDLC, batteries, catalyst supports, and field emission displays. By electrospinning of the PAN nanofiber and subsequent thermal treatments, CNFs has polarized electrodes in EDLC with a remarkable specific capacitance of ca. 297 F/g was obtained [167]. Seo et al. [137] showed that the ACNFs afforded good electronic conductivity, higher specific surface, suitable pore size, and higher content of surface oxygen functional groups. These unique properties of ACNFs were favorable for the diffusion of hydrated ions during charge/discharge within the electric double layers and more effective surface area was provided compared to CNFs, so On the bases of their high-power characteristics and excellent maintenance of specific capacitance, ACNFs could serve as useful electrode materials for supercapacitor applications.

8.16.4   REMOVAL OF POLLUTANTS

8.16.4.1   CNF AND ACNF FOR REMOVAL OF MICRO ORGANISM FROM WATER

Treatment processes for wastewater reuse and water treatment usually have adopted process such as biological treatment, coagulation, sand filtration, membrane filtration and activated carbon adsorption [168]. Recently, membrane filtration in water treatment has been used worldwide for reduction of particle concentration and natural organic material in water. Among the membrane processes, nanofiltration (NF) is the most recent technology, having many applications, especially for drinking water and wastewater treatment [169].

These nanofilters are reusable filters that have controlled porosity at the nanoscale and at the same time, can be formed into macroscopic structures with controlled geometric shapes, density, and dimensions. A carbonaceous nanofilter was fabricated by carbon nanofiber, in the best condition was used for MS2 virus removal. The results showed that at pressures of 8–11 bar the MS2 viruses were removed with a high efficiency by using the fabricated nanofilter. The results showed that the fabricated nanofilter had good water permeability, filtrates flux and could be used for virus removal with high efficiency [170].

8.16.4.2 CNF AND ACNF FOR REMOVAL OF VOLATILE ORGANIC COMPOUNDS (VOCS)

The manufacturers of specialty chemicals and pharmaceuticals generate effluent streams that contain trace amounts of aromatic and chlorinated hydrocarbons. Careful handling, recovery, or disposal of these toxic organics is one of the major environmental issues that confront such industries. Methods for the elimination of such contaminants from gaseous and liquid effluent streams are normally based on fixed-bed adsorption on carbonaceous materials. Traditionally, when recovery steps prove to be uneconomical or difficult, destruction of the organic contaminants is carried out by incineration. Such procedures, however, require elevated temperatures with associated high fuel costs. Catalytic degradation of organic contaminants into less toxic products may be an alternative low-temperature option. An effective catalyst is one on which the contaminant is initially strongly adsorbed and on which reaction with atomic species generated by the interaction of metallic components with the aqueous environment takes place [171]. VOCs (volatile organic compounds), such as toluene and benzene, are considered as pollutants [70]. Toluene is a hydrocarbon volatile organic compound with a low boiling point. This compound is quite harmful to human beings due to the easy conversion by other pollutants such as ozone and photochemical oxidants. Adsorption is a recommended method showing better control efficiency in removing toluene because it is emitted easily in low concentrations [20, 143].

There have been many attempts to reduce the level of pollution by VOCs using a variety of adsorption methods. Activated carbon nanofibers have attracted considerable attention as potential effective adsorbents for low concentrations of organic compounds by adsorption yield of 20–36%. Activated carbon fibers with diameters <10μm and pore sizes ranging from 8 to 20Å can be prepared using electrospinning of PAN solution [34]. However, despite the high specific surface area, the toluene adsorption capacity was limited due to the underdeveloped narrow micropores [172].

8.16.4.3 CNF AND ACNF FOR REMOVAL OF TOXIC MATERIALS IN ENVIRONMENT

Formaldehyde is one of the main pollutants in the atmosphere. In indoor air, formaldehyde mainly comes from decorating materials, paint, furniture glue, and chemical fiber carpets, and the concentration of formaldehyde is always relative low (<20 ppm). Even if the concentration of the formaldehyde is low, it can cause symptoms such as headache, nausea, coryza, pharyngitis, emphysema, lung cancer, and even death, so it is necessary to take effective measures for its removal. Adsorption by carbonaceous adsorbents is the most widely used method to purify the polluted air. Carbon nanostructure such a carbon nanofiber and fiber especially in activated form can be good adsorbents for formaldehyde [65]

Dichlorodiphenyltrichloroethane (DDT), DDT is a potential endocrine disruptor even at ng·L-1 levels. It is forbidden as a kind of pesticides from 1980s. While DDT is found in a higher concentration from the lake, river and the atmosphere, water, sediment, soil. It has been detected in many aquatic systems, from the Arctic Antarctic marine mammals to the birds, in the people's milk for human consumption, fish and so on. This raises serious problems in aquatic organisms and animals. Due to their harmful effects on the environment and biological body and the difficulty to degradation by the common treatment methods, it's important to use a suitable adsorbent to remove it activated carbon fiber and nanofiber are good adsorbent to eliminate it [173].

Another toxic material, which is harmful in environment is Arsenic. Arsenic contaminants in drinking water have been recognized as a serious environmental problem. Arsenic contamination in drinking water was found in the areas where water is extracted from groundwater with geological regions containing arsenic. But there are some cases of contamination from industries and mining as well. There are number of treatment methods for the removal of arsenic from water and wastewater. Chemical precipitation, ion exchange, ultra filtration, membrane techniques, lime softening and microbiological processes are the methods used for the treatment of water and wastewater containing arsenic. But the anaerobic process may be inhibited in chemical precipitation. Although reverse osmosis and ion exchange methods are effective in removing such pollutants, they are expensive in the operational procedure. These factors have limited the use of methods for the removal of arsenic and other toxic pollutants from water and wastewater especially in most of developing countries. Activated carbon nanofiber is successfully used to remove arsenic from wastewater [174].

8.16.5 CNF AND ACNF AS CATALYST SUPPORTS

In chemistry and biology, a carrier for catalyst is used to preserve high catalysis activity, increase the stability and life of the catalyst, and simplify the reaction process. An inert porous material with a large surface area and high permeability to reactants could be a promising candidate for efficient catalyst carriers [175]. Using an electrospun nanofiber mat as catalyst carrier, the extremely large surface could provide a huge number of active sites, thus enhancing the catalytic capability. The well-interconnected small pores in the nanofiber mat warrant effective interactions between the reactant and catalyst, which is valuable for continuous-flow chemical reactions or biological processes. Also, the catalyst can be grafted onto the electrospun nanofiber surface via surface coating or surface modification [163].

In the application of heterogeneous catalysts in liquid phase reactions, the rate of reaction as well as selectivity is often negatively influenced by mass transfer limitations in the stagnant liquid in the pores of the catalyst support [12]. Internal mass transfer limitations can be reduced by maximizing the porosity and lowering the tortuosity of the catalyst support. Particles and layers consisting of carbon nanofibers are promising catalyst supports because of the combination large pore volume (0.5–2 cm^3/g) and extremely open morphology, on one hand, and significant high surface area (100–200 m^2/g), on the other hand. The Scheme 6 compares the conventional and carbon nanofiber support catalysts (Fig. 8.15).

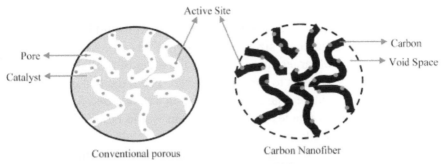

FIGURE 8.15 Carbon nanofiber as a catalyst support [11].

In order to maximize yield, in catalytic reactions, catalyst activity needs to be enhanced, this is typically achieved by developing catalytic sites having high intrinsic activities and by maximizing the number of active sites, e.g., by using high surface area support materials [175]. Carbon filaments are formed catalytically in metallic catalysts, particularly in Ni, Fe, and Co based catalysts, used for the conversion of carbon-containing gases, e.g., in steam reforming of hydrocarbons and Fischer-Tropsch synthesis. The carbon filament formation was detrimental for operation as they plugged reactors and deactivated catalysts.

Fiber type carbon nano materials, which are suitable as catalyst support can be classified into three types, namely, CNFs, CNTs, and single walled nanotubes (SWNTs). CNFs have a number of special characteristics that make them materials of promise as catalyst supports [11].

TABLE 8.8 Classification of CNF Characteristics for Catalyst Support Application

CNFs special characteristics for catalyst support application
Chemical stability for corrosive attack in acidic and basic environments
Having inert nature and with standing most organic solvents
Stability toward sintering and high-temperature gas reactions
Being conductive
Ability to apply in the field of electro catalysis such as fuel cell electrode

CNFs can be applied as catalyst supports in three ways:

- Using small aggregates of entangled nano fibers loaded with the catalytic active phase. The typical application would be in slurry reactors with aggregate sizes on the order of 10 cm.
- Application of larger aggregates (on the order of millimeters) of entangled CNF bodies to form a fixed bed. The fixed bed may be used as such in a single phase operation or as a trickle bed for gas-liquid operation. Advantages of such a system would be its high porosity and low tortuosity.
- The CNFs can form layers on structured materials such as foams, monoliths, or felts; this helps to keep diffusion distances short. The structured materials of choice obviously will also determine the hydrodynamic behavior of the reactor [8].

It appears rather easy to grow and attach CNFs on structured materials, that is, monoliths, graphite felts, silica fibers, metal filters, and metal foams. In many cases, excessive formation of CNFs weakens the support material and therefore the conditions of CNF formation needs to be accurately controlled [176].

CNFs appear to be well-attached to the supporting structures and a good explanation for this observation is still lacking. The layers of CNFs are indeed highly macroporous and should have low tortuosity, and the capability to allow fast mass and heat transfer has been demonstrated for hydrazine decomposition.

A comparable demonstration in liquid phase catalysis is lacking so far. In some cases, catalytic performance was claimed to be modified by the influence of the CNF-support material on either the metal particles or the mode of adsorption of reactants [12, 18].

Carbon nanofiber can be used as supporting agent for Pd catalysts. Pd catalysts have been applied to catalyze Heck reactions of various activated and nonactivated aryl substrates. The activity increased exponentially with a decrease in Pd particle size. The high surface area, mesoporous structure of carbon nanofiber and highly dispersed palladium species on carbon nanofibers makes up one of the most active and reusable heterogeneous catalysts for Heck coupling reactions. Pd nanoparticles supported on platelet CNFs appear to be an excellent catalyst due to high activity, low sensitivity towards oxygen, almost no or low issues with leaching and high stability in multicycles [175, 177] (Fig. 8.16).

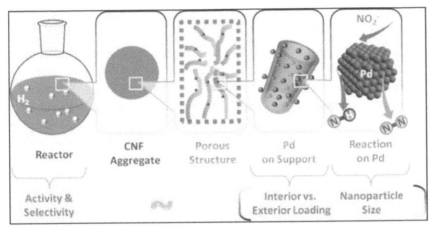

FIGURE 8.16 Carbon nanofiber as supporting agent for Pd catalysts.

8.16.6 CNF AS POLYMER REINFORCEMENT AGENTS

Early studies on electrospun based carbon nanofibers also included reinforcement of polymers. CNFs have an exceptional combination of mechanical and physical properties that make them ideal reinforcing materials for polymer composites. In order to properly incorporate CNFs into polymer composites, three major manufacturing challenges must be overcome:

- Dispersion of the CNFs in the matrix system,
- Uniform impregnation of the preform by the CNFs,
- Bonding and compatibility between the CNFs, matrix, and micro-sized reinforcement fibers [178].

As electrospun nanofiber mats have a large specific surface area and an irregular pore structure, mechanical interlocking among the nanofibers should occur. In general, the performance of a fibrous composite depends not only on the properties of the components, but also to a large degree on the coupling between the fiber and the matrix. In order to increase the internal laminar shear strength; numerous attempts have been made to improve bonding between the fiber and the matrix, consisting mostly of chemical and physical modification to the fiber surface [89].

Polymer/CNF nano composites can be prepared by different routes, including in situ polymerization, solution processing and melt mixing. The latter is the most common, given its simplicity and high yield, the compatibility with current industrial processes and the environmental advantage of a solvent-free procedure [179].

A novel approach in this field was the growing carbon nanofibers on the surface of conventional carbon fibers via carbon vapor deposition in the presence of a minuscule amount of metal catalyst [180]. The presence of the carbon nanostructures on the carbon fiber surface was found to enhance the surface area of perform from ~2 m²/g up to over 400 m2/g and consequently increased the interfacial bonding between the fiber and the matrix.

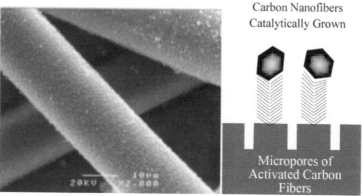

FIGURE 8.17　Carbon nanofiber on the surface of activated carbon fiber.

In order for a fiber to function successfully as reinforcement in high performance engineering materials, it must fulfill certain Criteria:

1. A small diameter with respect to its grain size is needed since there will be a low probability of intrinsic imperfections in the material and experimental evidence shows that the strength of the fiber increases as its diameter decreases. In this regard, carbon nanofibers would appear to be superior to other types of carbon fibers since their diameters [147], which are controlled by the size of the catalyst particles responsible for producing them, can be as low as 2 nm [81].

2. A high aspect ratio is needed which ensures that a very large fraction of the applied load will be transferred through the matrix to the stiff and strong fiber. In general, carbon nanofibers possess extremely small diameters and high aspect ratios, typically [81, 89, 90].

3. A very high degree of flexibility is desirable for the complex series of operations involved in composite fabrication. When carbon nanofibers are produced in a helical conformation they were found to possess appreciable elastic properties (Fig. 8.18).

FIGURE 8.18 Scanning electron micrograph showing the growth of carbon nanofibers on the surface of a bundle of a carbon fiber.

So because of the exceptional properties exhibited by carbon nanofibers such as their high tensile strength, modulus, and relatively low cost, such a material have a tremendous potential for reinforcement applications in its own right.

Thermoplastics such as polypropylene, polycarbonate, nylon, and thermo set such as epoxy, as well as thermoplastic elastomers such as butadiene-styrene di block copolymer, have been reinforced with carbon nanofibers for example. Carbon nanofibers with 0.5 wt% loading were dry-mixed with polypropylene powder by mechanical means, and extruded into filaments by using a single screw extruder. Decomposition temperature and tensile modulus and tensile strength have increased because of dispersion of CNF [121] (Fig. 8.19).

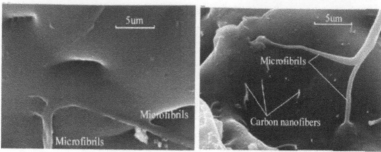

FIGURE 8.19 Fracture surface of neat and nanophased polypropylene (A: neat PP; B: CNF/PP) [181].

8.17 SUMMARY AND OUTLOOK

Carbon nanofibers have great advantages due to their high surface area to volume ratio. They have potential applications in the field of clean energy (solar cells, fuel cells and batteries), electronics, health (biomedical scaffolds, artificial organs), and environment (filter membranes). In summary, activated carbon nanofibers (ACNF) can be produced by different methods, which stabilizing, carbonizing and activating the electrospun PAN nanofiber is prefer among them.

However, few studies have applied stretching to electrospun precursor nanofibers before stabilization. Additionally, although aligned nanofibers with various degrees of alignment have been obtained by using various modified collecting devices, tension is rarely applied to the nanofiber assembly during stabilization to prevent shrinkage of the fibers and to ensure molecular orientations along the fiber axis to a large extent. Therefore, there is still considerable room for further improvement of the microstructural, electrical and mechanical properties of the final CNFs. Considering the increased interest in electrospinning and the wide range of potential applications for electrospun CNFs, commercialized products can be expected in the future.

Current advances in electrospinning technology provide important evidence of the potential roles in energy conversion and storage as well as water, and air treatment applications. Though electrospinning has become an essential technique for generating 1D nanostructures, the research exploration is still young, but promising, in energy applications. One of the drawbacks of PAN Electrospinning is that, it has been difficult to obtain uniform nanofibers with diameters below 50 nm using electrospinning. Another drawback is the relatively low production rate. In the near future, it is likely that research efforts will be focused on engineering the electrospinning process, with the ultimate goal of producing carbon nanofibers with diameters below 50 nm, and at a faster rate. The average diameter of ACNF was approximately 250 nm, ranging from 200 nm to 400 nm. The specific surface area and micropore volume of the carbon nanofibers (CNs) were 853 m²/g, and 0.280 cm³/g, respectively [34].

There is still a long way to go and much work to be done, and it is time to begin the task of turning the positive results born of research into the viable solutions born of engineering. In spite of a little information on the chemistry of activated carbon nanofiber the structural study is expected to provide significant insight on the structure itself and the formation of pores contributing to the surface area of carbon materials.

KEYWORDS

- **Activation**
- **Adsorption**
- **Carbon nanofiber**
- **Polyacrylonitrile**
- **Porosity**

REFERENCES

1. Zhang, Z., et al. (2009). Polyacrylonitrile and Carbon Nanofibers with Controllable. Nanoporous Structures by Electrospinning. *Macromolecules Material Engineering, 294,* 673–678.
2. Thavasi, V., Singh, G., & Ramakrishna, S. (2008). Electrospun Nanofibers in Energy and Environmental Applications. *Energy & Environmental Science, 1,* 205–221.
3. Fowlkes, J. D. (2008). Size-selectivity and Anomalous Subdiffusion of Nanoparticles through Carbon nanofiber-based membranes, *Nanotechnology 19,* 415301–415313.
4. Rożek, Z., Potential applications of nanofiber textile covered by carbon coatings. *Journal of Achievements in Materials and Manufacturing Engineering, 208(27),* 1.
5. Torres, M. et al. (2007). Hydrogen Adsorption by Nanostructured Carbons Synthesized by Chemical Activation, *Microporous and Mesoporous Materials, 98,* 89–93.

6. Strobel, R. et al. (2006). Hydrogen Storage by Carbon Materials. *Journal of Power Sources, 156,* 781–801.

7. Byung-Joo, K., Young-Seak, L., & Soo-Jin, P (2008). A Study on the Hydrogen Storage Capacity of Ni-plated Porous Carbon Nanofibers, International *Journal of Hydrogen Energy, 33,* 4112–4115.

8. Keane, M. A., & Park, C. (2003). Use of Carbon Nanofibers as Novel Support Material: Phenol Hydrogenation over Palladium *Catalical Community, 7.*

9. Sakoda, A., Nomura, T., & Suzuki, M. (1996). Activated Carbon Membrane for Water Treatments: Application to Decolorization of Coke Furnace Wastewater. *Adsorption, 3,* 93–98.

10. Jia-Zhia, W., (2008). Highly Sensitive Thin Film Sensor Based on Worm-like Carbon Nanofibers for Detection of Ammonia in Workplace. Chinese *Journal of Chemistry, 26,* 649–654.

11. Chinthaginjala, J., Seshan, K., & Lefferts, L. (2007). Preparation and Application of Carbon-Nanofiber Based Microstructured Materials as Catalyst Supports. Industrial *& Engineering Chemistry Research, 46,* 3968–3978.

12. Rodriguez, N. M., Kim, M. S., & Baker, R. T. K. (1994). Carbon Nanofibers: A Unique Catalyst Support Medium. *Journal of Physical Chemistry, 98,* 13108–13111.

13. Kim, C., & Lee, Y. H. (2003). EDLC Application of Carbon Nanofibers/Carbon Nanotubes Electrode Prepared by Electrospinning, in 203rd Meeting, Symposium Nanotubes, Nanoscale Materials, and Molecular Devices, The Electrochemical Society: Paris, France.

14. Kim, C. & Yang, K. (2003). Electrochemical Properties of Carbon Nanofiber Web as an Electrode for Supercapacitor Prepared by Electrospinning, *Applied Physics Letters, 83,* 20–26.

15. Metz, K. et al. (2006). Ultrahigh-Surface-Area Metallic Electrodes by Templated Electroless Deposition on Functionalized *Carbon Nanofiber Scaffolds Chemical Materials, 18,* 5398–5400.

16. McKnight, T. et al. (2003). Effects of Microfabrication Processing on the Electrochemistry of Carbon Nanofiber Electrodes. *Journal of Physical Chemistry, 107,* 10722–10728.

17. Huang, C., et al. (2007). Textural and electrochemical characterization of porous carbon nanofibers as electrodes for supercapacitors. *Journal of Power Sources, 172,* 460–467.

18. Sun, J., Wu, G., & Wang, Q. (2004). Adsorption Properties of Polyacrylonitrile-Based Activated Carbon Hollow Fiber, *Applied Polymer Science, 93,* 602–607.

19. Yu, Z. (2008). Pore Structure Analysis on Activated Carbon Fibers By Cluster and Watershed Transform Method, *Applied Surface Science.*

20. Oha, G.Y. et al. (2008). Adsorption of toluene on carbon nanofibers prepared by electrospinning. *Science of the Total Environment, 393,* 341–347.

21. Song, X., Wangb, C., & Zhang, D. (2009). Surface Structure and Adsorption Properties of Ultrafine Porous Carbon Fibers *Applied Surface Science, 255,* 4159–4163.

22. Quanming, L., & Wanxi, Z., (2009). Study on PAN-based Activated Carbon Fiber Prepared by KOH *Activation Method. Carbon, 66,* 70–75.

23. Reneker, D., & Chun, I. (1996). Nanometer Diameter Fibers of Polymer, produced by Electro spinning, *Nanotechnology, 7,* 216.

24. Ondarcuhu, T., & C. J (1998). Drawing a single nanofiber over hundreds of microns. *Europhysics Letters, 42,* 21220–21225.

25. Martin, C. (1996). Membrane-based Synthesis of Nanomaterials. *Chemistry of Materials, 8,* 1739–1746.

26. Ma, P., & Zhang, R. (1999). Synthetic Nano-Scale Fibrous Extracellular Matrix, *Journal of Biomedical Materials Research, 46,* 60–72.

27. Whitesides, G., & Grzybowski, B. (2002). Self-assembly at all scales. *Science, 295,* 2418–2421.

28. Nataraj, S. K., Yang, K. S., & Aminabhavi, T. M. (2011). Polyacrylonitrile-based nanofibers: A state-of-the-art review. *Progress in Polymer Science, 37,* 487–513.

29. Bhardwaj, N., & Kundu, S. C. (2010). Electrospinning: A fascinating fiber fabrication technique. *Biotechnology Advances, 28,* 325–347.

30. Ali, A., & El-Hamid, M. (2006). Electrospinning optimization for precursor carbon nanofibers. *Composites: Part A, 37,* 1681–1687.

31. Huang, Z. et al. (2003). A review on polymer nanofibers by electrospinning and their applications in nanocomposites. *Composites Science and Technology, 63,* 2223–2253.

32. Wang, C. et al. (2007). Electrospinning of Polyacrylonitrile Solutions at Elevated Temperatures, *Macromolecules, 40,* 7973–7983.

33. kim, J., Hong, I., & Lee, J. (1996). *Preparation of PAN based activated carbon fibers and its application for removal of sox in flue gas.*

34. Mitsubishi Rayon, Co., L. (2001). *Acrylonitrile-based precursor for carbon fiber and method for production thereof.*

35. Esrafilzadeh, D., Morshed, M., & Tavanai, H. (2009). An investigation on the stabilization of special polyacrylonitrile nanofibers as carbon or activated carbon nanofiber precursor. *Synthetic Metals, 159,* 267–272.

36. Im, J. et al. (2008). The Study of Controlling Pore Size on Electrospun Carbon Nanofibers for Hydrogen Adsorption. *Journal of Colloid and Interface Science, 318,* 42–49.

37. Kaixue, W. et al. (2009). Mesoporous Carbon Nanofibers for Supercapacitor Application. *Journal of Physical Chemistry C, 113,* 1093–1097.

38. Lee, S., Kimb, T., & Kimb, A. (2007). *Surface and Structure Modification of Carbon Nanofibers. Synthetic Metals, 157,* 644–650.

39. Li, C. et al. (2009). Porous Carbon Nanofibers Derived from Conducting Polymer: Synthesis and Application in Lithium-Ion Batteries with High-Rate Capability. *Journal of Physical Chemistry C, 113,* 13438–13442.

40. Ji, L., & Zhang, X. (2009). Manganese oxide nanoparticle-loaded porous carbon nanofibers as anode materials for high-performance lithium-ion batteries. *Electrochemistry Communications, 11,* 795–798.

41. Liu, C. et al. (2009). Preparation of Carbon Nanofibers through Electrospinning and Thermal Treatment. *Society of Chemical Industry, 58,* 1341–1349.

42. Maciá-Agulló, J. A. et al. (2007). Influence of carbon fibers crystallinities on their chemical activation by KOH and NaOH. *Microporous and Mesoporous Materials, 101,* 397–405.

43. Yusof, N., & Ismail, A. F. (2012). Post spinning and pyrolysis processes of polyacrylonitrile (PAN)-based carbon fiber and activated carbon fiber: A review. *Journal of Analytical and Applied Pyrolysis, 93,* 1–13.

44. Kim, C. et al. (2008*). Fabrications and Electrochemical Properties of Two-phase Activated Carbon Nanofibers from Electrospinning, Carbon, 42,* 34–39.

45. suzuki, p. (1994). *Actvated Carbon Fiber: Fundamentals and Applications. Carbon, 32,* 577–586.

46. Bajaj, P., Streekumar, T., & Sen, K. (2002). Structure Development during Dry-jet-wet Spinning of Acrylonitrile /Vinyl Acids and Acrylonitrile /Methyl Acrylate Copolymers. *Journal of Applied Polymer Science, 86,* 773–787.

47. Wilkinson, K. (2000). *Process for the preparation of carbon fiber.*

48. Panels, J., et al. (2008). Synthesis and characterization of magnetically active carbon nanofiber/iron oxide composites with hierarchical pore structures. *Nanotechnology, 19,* 455612–455619.

49. Wangxi, Z., Jie, L., & Gang, W. (2003). Evolution of Structure and Properties of PAN Precursors during their Conversion to Carbon Fibers. *Carbon, 41,* 2805–2812.

50. Zhou, Z. et al. (2009). Development of Carbon Nanofibers from Aligned Electrospun Polyacrylonitrile Nanofiber Bundles and Characterization of Their Microstructural, *Electrical, and Mechanical Properties. Polymer, 50,* 2999–3006.

51. Jagannathan, S. (2008). Structure and electrochemical properties of activated polyacrylonitrile based carbon fibers containing carbon nanotubes. *Journal of Power Sources, 185,* 676–684.

52. Aussawasathien, D. (2006). *Electrospun Conducting Nanofiber–Based Materials and their Characterization:* Effects of Fibers Characisctics on Properities and Applications, in the Graduate Faculty of the University of Akron. The University of Akronohio.

53. Fitzer, E. (1989). PAN-based Carbon-fibers Present State and Trend of the Technology from the Viewpoint of Possibilities and Limits to Influence and to Control the Fiber Properties by the Process parameters. *Carbon, 27,* 621–45.

54. Eslami Farsani, R. et al. (2009). FT-IR Study of Stabilized PAN Fibers for Fabrication of Carbon Fibers. World Academy of Science, *Engineering and Technology, 50,* 42–48.

55. Jimenez, V. et al. (2008). *Chemical Activation of Fish-bone Type Carbon Nanofiber.*

56. Lozano-Castello, D. (2001). Preparation of activated carbons from Spanish anthracite. I. Activation by KOH. *Carbon, 30,* 741–749.

57. Fu, R. et al. (2003). Studies on the Structure of Activated Carbon Fibers Activated by Phosphoric Acid. *Journal of Applied Polymer Science, 87,* 2253–2261.

58. Huidobro, A., Pastor, A. C., & Rodrıguez-Reinoso, F. (2001). Preparation of activated carbon cloth from viscous rayon. Part IV. *Chemical activation. Carbon, 30,* 389–398.

59. J. J. Purewal, et al. (2009). Pore size distribution and supercritical hydrogen adsorption in activated carbon fibers. *Nanotechnology, 20,* 204012 (6pp).

60. Macia-Agullo, J. A. et al. (2004) Activation of Coal Tar Pitch Carbon Fibres: Physical Activation vs. Chemical Activation. *Carbon, 42,* 1367–1370.

61. Guha, A. et al. (2001). Synthesis of Novel Platinum/Carbon Nanofiber Electrodes for Polymer Electrolyte Membrane (PEM) Fuel Cells. *Journal of Solid State Electrochemistry, 5,* 131–138.

62. Sun, J., & Wang, Q. (2005). Effects of the Oxidation Temperature on the Structure and Properties of Polyacrylonitrile-Based Activated Carbon Hollow Fiber. *Journal of Applied Polymer Science, 98,* 203–207.

63. Nguyen, T., & Bhatia, S. (2005). Characterization of activated carbon fibers using argon adsorption. *Carbon, 43,* 775–785.

64. Mochida, I., & Kawano, S. (1991). Capture of Ammonia by Active Carbon Fibers Further *Activated with Sulfuric Acid. Industrial & Engineering Chemistry Research 30,* 2322–2327.

65. Song, Y. et al. (2007). Removal of Formaldehyde at Low Concentration Using Various Activated Carbon Fibers. *Journal of Applied Polymer Science, 106,* 2151–2157.

66. Fitzer, E. (1986). *Carbon Fibers and Their Composites.* Berlin: Spiringer-Varlang 296p.

67. Casa-Lillo, M. et al. (2002). Hydrogen Storage in Activated Carbons and Activated Carbon Fibers. *Journal of Physical Chemistry part B, 106,* 10930–10934.

68. Purewal, J. J. (2009). Pore size distribution and supercritical hydrogen adsorption in activated carbon fibers. *Nanotechnology, 20,* 204012–204018.

69. Candy. Activated Carbon Fiber (ACF). 2002.

70. Singh, K. et al. (2002). Vapor-Phase Adsorption of Hexane and Benzene on Activated Carbon Fabric Cloth: Equilibria and Rate Studies. *Industrial & Engineering Chemistry Research, 41,* 2480–2486.

71. Brasquet, C., & Cloirec, P. (1994). Adsorption onto activated carbon fibers: Application to water and air treatments. *Carbon, 32,* 1307–1313.

72. Sun, J. et al. (2006). Effects of Activation Time on the Properties and Structure of Polyacrylonitrile-Based Activated Carbon Hollow Fiber. *Journal of Applied Polymer Science, 99,* 2565–2569.

73. Kim, B., Lee, Y., & Park, S. (2007). A Study on Pore-opening Behaviors of Graphite Nanofibers by a Chemical Activation Process. *Journal of Colloid and Interface Science, 306,* 454–458.

74. Jiuling, C., Qinghai, C., & Yongdan, L. (2006). Characterization and Adsorption Properities of Porous Carbon Nanofiber Granules. *China Particuology, 4,* 238–242.

75. Poole, C., & Owens, F. (2003). *Introduction to Nanotechnology,* Nalwa, H. S. ed., Hoboken, New Jersey: John Wiley & Sons, Inc. 396.

76. Dzenis, Y. (2004). *Spinning Continuous Fibers for Nanotechnology.* American Association for the Advancement of Science, 304, 1917–1919.

77. Ramakrishna, S. et al. (2005). *An Introduction to Electrospinning and Nanofibers.* Singapore: World Scientific Publishing Co. Pvt. Ltd.

78. Gu, S., Ren, J., & Vancso, G. (2005). Process optimization and empirical modeling for electrospun polyacrylonitrile (PAN) nanofiber precursor of carbon nanofibers. *European Polymer Journal, 41,* 2559–2568.

79. Kisin, E. R. et al. (2011). Genotoxicity of carbon nanofibers: Are they potentially more or less dangerous than carbon nanotubes or asbestos? *Toxicology and Applied Pharmacology, 252,* 1–10.

80. Park, C., & Keane, M. (2001). Controlled Growth of Highly Ordered Carbon Nanofibers from Y Zeolite Supported Nickel Catalysts. *Langmuir, 17,* 8386–8396.

81. Romero, A. et al. (2008). Synthesis and Structural Characteristics of Highly Graphitized Carbon Nanofibers Produced from the Catalytic Decomposition of Ethylene:

Influence of the Active Metal (Co, Ni, Fe) and the Zeolite Type Support. *Microporous and Mesoporous Materials, 110,* 318–329.

82. Jong, K. G. J. (2000). Carbon nanofibers: catalytic sythesize and application. *Catalysis Reviews, Sciences and Engineering, 42,* 481–510.

83. Park, C., & Baker, R. (1998). Catalytic Behavior of Graphite Nanofiber Supported Nickel Particles 2. The Influence of the Nanofiber Structure. *Journal of Physical Chemistry, 102,* 5168–5177.

84. Kima, J. et al. (2008). Preparation of polyacrylonitrile nanofibers as a precursor of carbon nanofibers by supercritical fluid process. *Journal of Supercritical Fluids, 47,* 103–107.

85. Wu, S. et al. (2008). Preparation of PAN-based Carbon Nanofibers by Hot-stretching. *Composite Interfaces, 15,* 671–677.

86. Barranco, V. et al. (2010). Amorphous Carbon Nanofibers and Their Activated Carbon Nanofibers as Supercapacitor Electrodes. *Journal of Physical Chemistry C, 114,* 10302–10307.

87. Yoon, S. et al. (2004). Carbon Nano-rod as a Structural Unit of Carbon Nanofibers. *Carbon, 42,* 3087–3095.

88. ZhEng., R. et al. (2006). Preparation, Characterization and Growth Mechanism of Platelet Carbon Nanofibers. *Carbon, 44,* 742–746.

89. Rodriguez, N. M. (1993). A Review of Catalytically grown Carbon Nanoflbers. *Commentaries and Reviews, 8,* 12–19.

90. Kim, C. et al. (2004). Raman Spectroscopic Evaluation of Polyacrylonitrile-based Carbon Nanofibers Prepared by Electrospinning. *Journal of Raman Spectroscopy, 35,* 928–933.

91. Lingaiah, S. et al. (2005). *Polyacrylonitrile-Based Carbon Nanofibers Prepared by Electrospinning.*

92. Kurban, Z. et al. (2000). *Graphitic nanofibers from electrospun solutions of PAN in dimethylsulphoxide.*

93. Chun, I. et al. (1999). Carbon nanofibers from polyacrylonitrile and mesophase pitch. *Journal of Advanced Materials 31,* 36–41.

94. Wang, Y., Serrano, S., & Aviles, J. (2002). Conductivity measurement of electrospun PAN-based carbon nanofiber. *Journal of Materials Science Letters, 21,* 1055–1057.

95. Panapoy, M., Dankeaw, A., & Ksapabutr, B. (2008). Electrical Conductivity of PAN-based Carbon Nanofibers Prepared by Electrospinning Method. *Thammasat Int. J. Sc. Tech, 13,* 88–93.

96. Hou, H., & Reneker, D. (2004). *Carbon Nanotubes on Carbon Nanofibers*: A Novel Structure Based on Electrospun Polymer Nanofibers. Advanced Materials, *16,* 69–73.

97. Kim, S. & Lee, K. (2004). Carbon Nanofiber Composites for the Electrodes of Electrochemical Capacitors. *Chemical PhysicsLetters, 400,* 253–257.

98. Jalili, R., Morshed, M., & Hosseini Ravandi, S. (2006). Fundamental Parameters Affecting Electrospinning of PAN Nanofibers as Uniaxially Aligned Fibers. *Journal of Applied Polymer Science, 101,* 4350–4357.

99. Wang, Y., Serrano, S., & Santiago-Avile,' J. (2003). Raman characterization of carbon nanofibers prepared using electrospinning. *Synthetic Metals, 138,* 423–427.

100. Ch. Kim, et al. *(2004)*. Raman spectroscopic evaluation of polyacrylonitrile-based carbon nanofibers prepared by electrospinning. *JOURNAL OF RAMAN SPECTROS-COPY, 35*, 928–933.

101. Jalili, R., Morshed, M., & Ravandi, S. H. (2006). Fundamental Parameters Affecting Electrospinning of PAN Nanofibers as Uniaxially Aligned Fibers. *Journal of Applied Polymer Science, 101,* 4350–4357.

102. Ch. Wang, et al. (2007). Electrospinning of Polyacrylonitrile Solutions at Elevated Temperatures. *Macromolecules, 40,* 7973–7983.

103. Zussman, E. et al. (2005). Mechanical and Structural Characterization of Electrospun PAN-derived Carbon Nanofibers. *Carbon, 43,* 2175–2185.

104. Ci, L. et al. (2001). Carbon Nanofibers and Single-walled Carbon Nanotubes Prepared by the Floating Catalyst Method. *Carbon, 39,* 329–335.

105. Kalayci, V. E. (2005). Charge consequences in electrospun polyacrylonitrile (PAN) nanofiber. *Polymer, 46,* 7191–7200.

106. Blackman, J. et al. (2006). Activation of Carbon Nanofibers for Hydrogen Storage. *Carbon, 44,* 44–48.

107. Yördem, O., Papila, M., & Menceloglu, Y. (2001). Prediction of Electrospinning Parameters for Targeted Nanofiber Diameter.

108. Vicky, V., Katerina, T., & Niko, s.C. (2006). Carbon Nanofiber-Based Glucose Biosensor. *Analytical Chemistry, 78,* 5538–5542.

109. Deitzel, J. M. et al. (2001). The Effect of Processing Variables on the Morphology of Electrospun Nanofibers and Textiles. *Polymer, 42,* 261–272.

110. Boland, E. D. (2001). Taloring Tissue Engineering Scaffold Using Electrostatic Processing Techniques: A Study of Poly (Glycolic Acid) Electrospinning. *Journal of Macromolecular Science Pure and Applied Chemistry 12,* 1231–1243.

111. Ryu, Y. J. (2003). Transport Properties of Electrospun Nylon 6 Nonwoven Mats. *European Polymer Journal, 39,* 1883–1889.

112. Katti, D. S. ct al. (2004). Bioresorbable nanofiber-based systems for wound healing and drug delivery: optimization of fabrication parameters. *Journal of Biomed Mater. Res B Appl Biomater, 70,* 286–296.

113. Reneker, D., & Chun, I. (1996). Nanometer Diameter Fibers of Polymer, Produced by Electrospinning. *Nanotechnology, 7,* 216.

114. Demir, M. M. (2003). Electrospining of Polyurethane Fibers. *Polymer, 43,* 3303–3309.

115. Sukigara, S. et al. (2004). Regeneration of Bombyx Mori Silk by Electrospinning. Part 2. Process Optimization and Empirical Modeling Using Response Surface-Methodology, *Polymer, 45,* 3701–3708.

116. He, J., Wan, Y., & Yu, J. (2008). Effect of Concentration on Electrospun Polyacrylonitrile (PAN) Nanofibers. *Fibers and Polymers, 9,* 140–142.

117. Agend, F., Naderi, N., & Alamdari, R. (2007). Fabrication and Electrical Characterization of Electro spun Polyacrylonitrile-Derived Carbon Nan fibers Farima. *Journal of Applied Polymer Science, 106,* 255–259.

118. Jalili, R., Hosseini, A., & Morshed, M. (2005). The Effects of Operating Parameters on the Morphology of Electrospun Poly acrilonitrile Nanofibers. *Iranian Polymer Journal, 14,* 1074–1081.

119. Ma, G., Yanga, D., & Nie, J. (2009). Preparation of Porous Ultrafine Polyacrylonitrile (PAN) Fibers by Electrospinning. *Polymer Advanced Technology, 20,* 147–150.

120. Baker, S., et al. (2006). Functionalized Vertically Aligned Carbon Nanofibers as Scaffolds for Immobilization and Electrochemical Detection of Redox-Active Proteins. *Chemistry of Materials, 18,* 4415–4422.

121. Hasan, M., Zhou, Y., & Jeelani, S. (2006). Thermal and Tensile Properties of Aligned Carbon Nanofiber Reinforced Polypropylene. *Materials Letters, 61,* 1134–1136.

122. Chenga, J. et al. (2004). Long Bundles of Aligned Carbon Nanofibers Obtained by Vertical Floating Catalyst Method. *Materials Chemistry and Physics, 87,* 241–245.

123. Weisenbergera, M. et al. (2009). The Effect of Graphitization Temperature on the Structure of Helical-ribbon Carbon Nanofibers. *Carbon, 47,* 2211–2218.

124. Zhang, L., & Hsieh, Y. (2009). Carbon Nanofiberswith Nanoporosity and Hollow Channels from Binary Polyacrylonitrile Systems. *European Polymer Journal, 45,* 47–56.

125. Wang, Y., & Santiago, J. (2002). Early Stages on the Graphitization of Electro statically Generated PAN Nanofibers in Conference on Nanotechnology (IEEE-NANO 2002). Pensylvania, USA, 29–32.

126. Wang, Y., Serrano, S., & Santiago-Avile, J. (2003). Raman Characterization of Carbon Nanofibers Prepared Using Electrospinning. *Synthetic Metals, 138,* 423–427.

127. Ko, T., Kuo, W., & Hu, C. (2001). Raman Spectroscopic Study of Effect of Steam and Carbon Dioxide Activation on Microstructure of Polyacrylonitrile- Based Activated Carbon Fabrics. *Journal of Applied Polymer Science, 81,* 1090–1099.

128. Yamashita, Y. et al. (2008). Carbonization Conditions for Electrospun Nanofiber of Polyacrylonitrile Copolymer. *Indian Journal of Fiber and Textile Research, 33,* 345–353.

129. Su, C. et al. (2012). *PAN-based Carbon Nanofiber Absorbents Prepared Using Electrospinning. Fibers and Polymers, 13,* 436–442.

130. Shimada, I., & Takahagi, T. (1986). FT-IR Study of the Stabilization Reaction of Polyacrylonitrile in the Production of Carbon Fibers. *Journal of Polymer Science, Part A: Polymer Chemistry, 24,* 1989–1995.

131. Coleman, M., & Petcavich, R. (1987). Fourier Transform Infrared Studies on the Thermal Degradation of Polyacrylonitrile. *Journal of Polymer Science, Part A: Polymer Chemistry, 16,* 821–832.

132. Zhang, S. et al. (2008). Structure evolution and optimization in the fabrication of PVA-based activated carbon fibers. *Journal of Colloid and Interface Science, 321,* 96–102.

133. Wang, P., Hong, K., & Zhu, Q. (1996). Surface Analyzes of Polyacrylonitrile-Based Activated Carbon Fibers by X-ray Photoelectron Spectroscopy. *Journal of Applied Polymer Science, 62,* 1987–1991.

134. Yoon, S. et al. (2004). KOH Activation of Carbon Nano fibers. *Carbon, 42,* 1723–1729.

135. Jim'enez, V. et al. (2009). Microporosity Development of Herringbone Carbon Nanofibers by RbOH Chemical Activation. *Research Letters in Nanotechnology, 5,* 52–56.

136. Tavanai, H., Jalili, R., & Morshed, M. (2009). Effects of Fiber Diameter and CO_2 Activation Temperature on the Pore Characteristics of Polyacrylonitrile Based Activated Carbon Nanofibers. *Surface and Interface Analysis, 41,* 814–819.

137. Seo, M., & Park, S. (2009). Electrochemical Characteristics of Activated Carbon Nanofiber Electrodes for Super capacitors. *Materials Science and Engineering, 164,* 106–111.
138. Im, J., Jang J., & Lee, Y. (2009). Synthesis and Characterization of Mesoporous Electrospun Carbon Fibers Derived From Silica Template. *Journal of Industrial and Engineering Chemistry, 15,* 914–918.
139. Lillo-Ro'denas, M., Cazorla-Amoro's, D., & Linares-Solano, A. (2003). Understanding Chemical Reactions Between Carbons and NaOH and KOH An Insight into the Chemical Activation Mechanism. *Carbon, 41,* 267–275.
140. Chuang, C. et al. (2008). Temperature and Substrate Dependence of Structure and Growth Mechanism of Carbon Nanofiber. *Applied Surface Science, 254,* 4681–4687.
141. Ji, L., & Zhang, X. (2009). Electrospun Carbon Nanofibers Containing Silicon Particles as an Energy-Storage Medium. *Carbon, 47,* 3219–3226.
142. Kim, C., et al., (2006). Fabrication of Electrospinning-Derived Carbon Nanofiber Webs for the Anode Material of Lithium-Ion Secondary Batteries. *Advanced Functional Materials, 16: p.* 2393–2397.
143. Oh, G., et al., (2008) Preparation of the novel manganese-embedded PAN-based activated carbon nanofibers by electrospinning and their toluene adsorption. *Journal of Analytical applied pyrolysis,. 81:* p. 211–217.
144. Vasiliev, L., et al., (2007). Hydrogen Storage System Based on Novel Carbon Materials and Heat Pipe Heat Exchanger. International *Journal of Thermal Sciences, 46: p.* 914–925.
145. Ahn, C., et al., (1998). Hydrogen Desorption and Adsorption Measurements on Graphite Nanofibers. *Applied Physics Letters, 73:* p. 77–81.
146. Tibbetts, G., Meisner G, &. Olk C, (2001). Hydrogen Storage Capacity of Carbon Nanotubes, Filaments, and Vapor-grown Fibers. *Carbon, 39,* 2291–2301.
147. Sharon, M., et al., (2004). Synthesis of Carbon Nano-Fiber from Ethanol and It's Hydrogen Adsorption Capacity. *Carbon,. 61:* p. 21–26.
148. Chahine, R. &. Bénard, P (2001): Assessment of Hydrogen Storage on Different Carbons, in Metal Hydrides and Carbon for Hydrogen Storage Richard Chahine (Canada).
149. Browning, D., et al., (2002). Studies into the Storage of Hydrogen in Carbon Nanofibers: *Proposal of a Possible Reaction Mechanism. Nano letters, 2: p.* 201–205.
150. Rzepka, M., et al., (2005) Hydrogen Storage Capacity of Catalytically Grown Carbon Nanofibers. *Journal of Physical Chemistry C, 109,* 14979–14989.
151. Byung-Joo, K., Young-Seak L., &. Soo-Jin, P (2008). Preparation of Platinum-decorated Porous Graphite Nanofibers, and their Hydrogen Storage Behaviors. *Journal of Colloid and Interface Science, 318,* 530–533.
152. Cook, B.,. (2001): An Introduction to Fuel Cells and Hydrogen Technology *Vancouver.*
153. Zeches, R. (2002). Carbon Nanofibers as a Hydrogen Storage Medium for Fuel Cell Applications in the Transportation Sector.
154. Kim, D., et al., (2005). Electrospun Polyacrylonitrile-Based Carbon Nanofibers and Their Hydrogen Storages. *Macromolecular Research, 13,* 521–528.

155. Ghasemi, M., et al., (2011). Activated carbon nanofibers as an alternative cathode catalyst to platinum in a two-chamber microbial fuel cell. International *Journal of Hydrogen Energy, 36*, 13746–13752.

156. Lee, Y., & Sun Im, J. (2010). Preparation of Functionalized Nano fiber sand their Applications, in Nanofibers, Kumar A, Editor.

157. Tran, P., Zhang, L., & Webster, T. (2009). Carbon Nanofibers and Carbon Nanotubes in Regenerative Medicine. *Advanced Drug Delivery Reviews, 61*, 1097–1114.

158. Seidlits, S. K., Lee, J. Y., & Schmidt, C. E. (2008). Nanostructured scaffolds for neural applications. *Nanomedicine, 3*, 183–199.

159. Tavangarian, F., & Li, Y. (2012). Carbon nanostructures as nerve scaffolds for repairing large gaps in severed nerves. *Ceramics International, 38(8)*, 6075–6090.

160. Ma, C. et al. (2012). Phenolic-based carbon nanofiber webs prepared by electrospinning for supercapacitors. *Materials Letters, 76*, 211–214.

161. Lipka, S. (1998). Carbon Nanofibers and Their Applications for Energy Storage, in Battery Conference on Applications and Advances, I. X. Library, Editor. Long Beach, CA USA, 373–374.

162. Ji, L. et al. (2009). Porous Carbon Nanofibers from Electrospun Polyacrylonitrile SiO2 Composites as an Energy Storage Material. *Carbon, 47*, 3346–3354.

163. Jian, F. et al. (2008). *Applications of Electrospun Nanofibers Chinese Science Bulletin, 53*, 2265–2286.

164. Mukherjee, R. et al. (2012). Nanostructured electrodes for high-power lithium ion batteries. *Nano Energy, 1*, 518–533.

165. Donough, J. et al. (2009). Carbon nanofiber supercapacitors with large areal capacitances. *Applied Physics letters, 95*, 243109–243111.

166. Su, Y. et al. (2009). Activation of Ultra-Thin Activated Carbon Fibers as Electrodes for High Performance ElectrochemicalDouble Layer Capacitors. *Journal of Applied Polymer Science, 111*, 1615–1623.

167. Tao, X. et al. (2006). Synthesis of Multi-branched Porous Carbon Nanofibers and their Application in Electrochemical Double-layer Capacitors. *Carbon, 44*, 1425–1428.

168. Bae, S. D. et al. (2007). Preparation, characterization, and application of activated carbon membrane with carbon whiskers. *Desalination, 202*, 247–252.

169. Tahaikt, M. et al. (2007). Fluoride removal from groundwater by nanofiltration. *Desalination, 212*, 46–53.

170. Mostafavia, S., Mehrnia, M., & Rashidi, A, (2009). Preparation of Nanofilter from Carbon Nanotubes for Application in Virus Removal from Water. *Desalination, 238*, 88–94.

171. Park, C. et al. (2000). Use of Carbon Nanofibers in theRemoval of Organic Solvents from Water. *Langmuir, 16*, 8050–8056.

172. Cuervo, M. et al. (2008). Effect of Carbon Nanofiber Functionalization on the Adsorption Properties of Volatile Organic Compounds. *Journal of Chromatography, 1188*, 264–273.

173. Guo, Y. et al. (2009). Adsorption of DDT by Activated Carbon Fiber Electrode, in International Conference on Energy and Environment Technology.

174. Lee, S. (2010). Application of Activated Carbon Fiber (ACF) for Arsenic Removal in Aqueous Solution. Korean, *Journal of Chemical Engineering 27*, 110–115.

175. Shuai, D. et al. (2012). Enhanced Activity and Selectivity of Carbon Nanofiber Supported Pd *Catalysts for Nitrite Reduction. Environmental Science & Technology, 46,* 2847–2855.

176. Coelho, N. et al. (2008). Carbon Nanofibers: a Versatile Catalytic Support. *Materials Research Society, 11,* 353–357.

177. Zhu, J. et al. (2009). Carbon Nanofiber-supported Palladium Nanoparticles as Potential Recyclable Catalysts for the Heck Reaction. *Applied Catalysis A: General, 352,* 243–250.

178. Rodriguez, A. et al. (2011). Mechanical properties of carbon nanofiber/fiber-reinforced hierarchical polymer composites manufactured with multi scale-reinforcement fabrics. *Carbon, 49,* 937–948.

179. Novais, R., Covas, J., & Paiva, M. (2012). The effect of flow type and chemical functionalization on the dispersion of carbon nanofiber agglomerates in polypropylene. *Composites: Part A, 43,* 833–841.

180. Lim, S., et al. (2004). Surface Control of Activated Carbon Fiber by Growth of Carbon Nanofiber *Langmuir, 2004,* 5559–5563.

181. Hasan, M. M., Zhou, Y., & Jeelani, S. (2006). Thermal and tensile properties of aligned carbon nanofiber reinforced polypropylene. *Materials Letters, 61,* 1134–136.

INDEX